DATE DUE

Just
from

Just Seconds from the Ocean

Coastal Living in the Wake of Katrina

William Sargent

University Press of New England

Hanover and London

Published by University Press of New England,
One Court Street, Lebanon, NH 03766
www.upne.com
© 2007 by University Press of New England
Printed in the United States of America
5 4 3 2 1

Library of Congress Cataloging-in-Publication Data

Sargent, William, 1946–
Just seconds from the ocean : coastal living in the
wake of Katrina / William Sargent.
p. cm.
Includes bibliographical references and index.
ISBN-13: 978-1-58465-689-0 (cloth : alk. paper)
ISBN-10: 1-58465-689-1 (cloth : alk. paper)
1. Coast changes—United States. 2. Coastal zone management—
United States. 3. Hurricane Katrina, 2005—Environmental
aspects. 4. Hurricanes—United States. I. Title.
GB460.A2S27 2007
363.34'922097309146—dc22   2007041721

This book is dedicated to the people of New Orleans.

Contents

CONTENTS

Preface

The world changed on August 29, 2005. Overnight, the inhabitants of New Orleans experienced the effects of history's first major global warming disaster. The next day, people in other vulnerable coastal cities started to think the unthinkable: What would happen to New York, Miami, or Boston in the wake of another Hurricane Katrina?

However, the disaster also initiated a national debate about flood control and coastal policy. It was not a clear-cut process. Crisis management was the order of the day. In New Orleans, people had to be cared for, levees had to be rebuilt, and food, water, electricity, and sewer infrastructure all had to be restored. But while these many short-term decisions were being made, they were having long-term impacts on future decisions. This was policymaking on the fly. I wanted to see it up close.

My first chance came immediately after Katrina when the *Cape Cod Times* asked me to write an op-ed piece about rebuilding New Orleans. The papers were full of editorials saying we should spend whatever it cost to rebuild New Orleans exactly as it had been before. I simply pointed out that it hadn't made very much sense to build a city twenty feet below sea level in the first place.[1] It was pretty easy to sit up in Massachusetts and make such broad pronouncements about all the mistakes Louisiana had made in the past three hundred years. It was also easy to make the point that sea-level rise and future hurricanes made the long-term future of New Orleans bleak indeed. But as I wrote, I also realized that the issue was far more complicated than could be conveyed in a simple op-ed article. I decided to spend a year traveling to specific hot spots up and down the East and Gulf Coasts to investigate how past and future storms would affect our shores.

In preparation for my trip I continued to monitor developments in New Orleans. When federal officials decided to only rebuild levees strong enough to withstand a Category 3 hurricane instead a Category 5 hurricane, they made the tacit decision that as a nation we were not going to protect every vulnerable city against a Category 5 hurricane. This made economic sense. It would be simply too expensive to build Category 5 systems to protect every major coastal city. This was a partial repudiation of the hard-structure, build-whatever-you-have-to school of engineering.

On the other hand, when a panel of nationally recognized experts presented a plan that suggested that certain areas of New Orleans not be rebuilt, it was dead on arrival. Even the mayor who had helped set up the panel walked away from their well-thought-out recommendations. This was a repudiation of the human-behavior-should-be-managed-to-fit-the-environment school of thought.

The results should not have surprised me. I had spent several years writing about similar phenomena on Cape Cod. Psychologists tell us that humans have an innate, almost biological attachment to their homes. People will fight to the death to save their houses and if they lose them will suffer almost as much as if they had lost a spouse or had been kicked out of a job. Interestingly enough, psychologists also discovered that people can accept it if they feel they lost their houses because of a hurricane but suffer more if they believe they lost their homes because of human intervention like zoning. But the loss of a home doesn't stop there; it continues to reverberate through families for generations. Of course, in New Orleans many people lost all three psychological props: their homes, their jobs, and their families.[2]

But I also realized that something else was going on. I had originally thought that people in other coastal areas would learn lessons from New Orleans, but not that long-term policy would actually be made for them in the caldron of crisis swirling about the stricken city. I also learned that this is how it has always been. Because Congress decides on the budget for the Army Corps of Engineers, and because almost three quarters of the Corps' budget goes toward protecting New Orleans and keeping the Mississippi River navigable, the decisions that Congress makes about New Orleans automatically set policy for the rest of the nation. When the nation decided to adopt the "levees only" policy after the disastrous Mississippi floods in 1927, the Army Corps of Engineers followed suit by starting to build massive seawalls along large stretches of our nation's coasts. That also led to the aphorism that while New Orleans wouldn't be possible without the Army Corps of Engineers, the Army Corps of Engineers wouldn't be possible without New Orleans.

So this book is the result of spending a year traveling to different locations to investigate specific coastal issues. Chapter 1 starts with the problem: coastal erosion, global warming, and hurricanes. Chapter 2 looks at what happened in New Orleans in some detail. Chapter 3 investigates the history of coastal development up and down the East and Gulf Coasts and shows how Hurricane Katrina will affect the lives of people who wish to continue living on the coast. Chapter 4 looks at the future of both New Orleans and the hurricane-prone East and Gulf Coasts.

Today, I find myself guardedly optimistic. After a year of crisis management, introspection, and accepting blame for the failures of hard engineering in New Orleans, the Army Corps of Engineers is on the verge of adopting a policy that will concentrate on enhancing the natural redundancy of the marshes, barrier islands, and natural ridges that have always protected the region. This represents a marked turnabout from the Corps' former reliance on hard engineering solutions. It happened mostly out of the limelight, after the media's attention had turned elsewhere, but it will have reverberations throughout the country. It remains to be seen whether this will be enough to save New Orleans, which has suffered an almost lethal blow.

Along with the realization that hard engineering solutions alone would not work to protect other coastal areas came the parallel realization that people can no longer continue to live wherever they want. Rivers have to have room to flood, and barrier beaches must be able to move to protect the mainland. At the same time the Army Corps of Engineers was about to hire a firm to use these new biological engineering principles to prevent floods in New Orleans, the National Academy of Sciences released a report that recommended that states replace local communities in making long-term decisions about coastal development throughout the nation.[3] In essence it advocated managing human behavior and suggested that this could be accomplished more effectively from a higher level than strictly on a local basis.

All across the globe, people seem poised to finally do something about climate change. Farmers are planting corn for their own self-financed ethanol facilities, towns are building wind turbines, companies are developing new technologies, and voters are demanding that government mandate higher mileage cars. After governmental denial and wishful thinking for almost two decades, the gears of a new paradigm are shifting into place, and the entrepreneurial drive of our still innovative nation is about to be unleashed.

In Europe, the widely touted Stern report has shifted the argument about climate change away from the so-called controversy about the science of global warming to a relatively easy economic solution. In the six-hundred-page report, Sir Nicholas Stern, former economist for the World Bank, argues that if the world expends just 1 percent of its economic output to reduce carbon emissions now, it can prevent up to 20 percent in economic losses that will occur within the next one hundred years. This would be the equivalent of all the economic losses suffered in both World Wars plus the 1929 depression.[4]

The Stern report is aimed largely at the United States, the world's prime rogue nation for refusing to sign the Kyoto Global Warming Treaty. An October 31, 2006, editorial in the London *Financial Times* grumbled that "The Bush Administration remains infuriatingly relaxed about climate change because the U.S. mainland is likely to remain relatively untouched by rising sea levels."[5] That might be an easy charge to level against the United States if you are sitting in London behind the largely untested Thames storm surge barrier. But if you are sitting anywhere on the hurricane-prone East or Gulf Coasts of the United States, you can relate to the world's pique at the Bush Administration's denial about the reality of global warming. Thanks to Hurricane Katrina, we now know we are living on the edge of a vulnerable coast, surviving on the basis of borrowed time.

W.S.

Ipswich, Massachusetts

Just Seconds
from the Ocean

Introduction

A Childhood Memory—Monomoy Island

It is late afternoon. Our government-issue jeep is loaded with clam rakes and fishing rods, olive oil and kerosene—the essential necessities for my family's annual fishing and culinary expedition to Monomoy Island on the southernmost tip of Cape Cod. We hold our breath as my father inches our jeep up two narrow planks and onto the rickety barge that serves as the only ferry to the nearby island. A jeep is the only vehicle that can fit on this homemade platform. The barge owner wraps a piece of old frayed cord around the starting wheel and pulls—nothing. He cusses and pulls again. Swearing seems to help; the motor pops and sputters into life, and we move slowly over the shallow waters.

Just before landing we see the rusty hulks of several old Model Ts mired in the sand. My father eases the jeep back down the narrow planks, and we deflate the tires to avoid a similar fate. Slowly we start to lurch and chug down the sandy ruts that run the length of this eight-mile-long island. Our greatest thrill is to sit on the tailgate, where my sisters and I can feel the sand spinning off the tires and the heat of the exhaust singeing our naked legs. The jeep moves so slowly we can jump off, land on our backsides, run a few steps, and still have enough time to leap back on the moving tailgate. We are told to stop our shenanigans but continue to pretend to be thrown off on every bounce, and gales of laughter give us away. Other times we are told to jump out and shove boards under the tires to give the jeep traction through the sandy dunes.

It is easier once we make it to the foreshore. Here all we have to do is swerve up and down the beach to avoid the incoming breakers, which always seem larger on the island. It is usually dark before we reach the far end of Monomoy. Occasionally we see the headlights of another beach buggy lurching back and forth through the dunes, but mostly we are alone, save for the ghostly apparitions of beach toads trying to scramble out the ruts as the bouncing beam of our headlights bears down upon them.

The tracks are as firm and smooth as asphalt on the backside of the island. Here flocks of shorebirds huddle in the marsh grass, and horseshoe crabs mate along the shore. Finally we see what we are looking for, the low-lying silhouettes of a handful of lonely dark shacks huddled around a sandy cove. Our shack is a casual affair with an outhouse, kerosene lanterns, and thin wooden walls. It is my job to prime the hand pump with a bucketful of water left by a thoughtful former occupant. Soon sweet fresh water is gushing into the tin basin. After the kerosene is poured and the lanterns are lit we spread our sleeping bags onto simple wooden bunks and fall into a deep slumber. The only light is the long beam of the Monomoy Light House that sweeps over the dunes, into our windows, and out again over the black Atlantic.

We wake up early to be on the point before sunrise. Monomoy Point is a place of shoals and sandbars, rips and runnels. This is where cold, green waters of the Labrador Current mix with the warm waters of the nearby Gulf Stream. This is where the twelve-foot tides of the North Atlantic flow into the three-foot tides of the mid-Atlantic. This is where the combined energy of wind and water delivers the millions of tons of sand that erode off the thirty miles of continuous beach and bluffs that run from Truro to Chatham. Here the sand piles up in treacherous shoals and sandbars. Here the sandbars migrate onto the beach, where the wind fashions them back into thirty-foot sand dunes. The dunelands stretch almost two miles thick from the ocean to the backside, making this southernmost tip of Monomoy look like the head of a giant Indian war club with a long narrow handle.

We cast our jigs into the maelstrom and bounce them back along the bottom. The currents are so strong we can't reel our lures against the tide. A large striped bass lunges out of a nearby hole and terns gather raucously overhead. We can see bluefish bursting through the green wall of a current to attack a school of squid. The water turns reddish black with the squids' ejected ink. Seals gather to harass the bluefish, and where there are seals, can sharks be far behind? Our neighbor swears he once saw two great white sharks continuously beaching themselves on Monomoy to trap menhaden against the shore. After each lunge the sharks would have to flip around like giant minnows to make it back into the water.

The bluefish come in fast and furiously, until the tide slackens and we change bait to catch some fluke. I walk into the dunes to nap in a patch of sweet-smelling beach peas. By late afternoon we return to the Powder Hole to dig some clams and watch my father clean our fish on a simple wooden cleaning table built into the side of the shack.

We hear my mother pouring olive oil into a hot frying pan and run inside to help her dip the fillets into cornmeal and watch their edges curl as they hit the hot oil. Soon the entire shack fills with the smell of steamed clams, sizzling fish, hot butter, and cold beer.

If there were ever a place to build a beach house, the tip of Monomoy Island would be the place. You could build your house amid rolling acres of protective sand dunes overlooking the wild Atlantic. But this is not where early Cape Codders wanted to be. After a day spent rocking about in a narrow dory, the last thing they wanted to see was more ocean. They preferred to huddle together around the Powder Hole. More than two hundred fishermen once lived around this shallow protected cove in a community called Whitewash, a small, sandy village complete with its own church and public school.

People started settling on this narrow sandbar as early as 1711, when an enterprising Chathamite established a tavern on the backside of Monomoy so fishermen could nip in for the night and be out the next morning to save time. Their boats would lie at anchor in the nearby Wreck Cove. The cove was named for both the many ships that grounded on Monomoy's treacherous shoals and the many wreckers who attracted them with false beacons. Locally the wreckers were called mooncussers because they could not cause shipwrecks when the moon was full. It is reported that when a local minister looked out over the bowed heads of his congregation and saw a ship about to founder, he told his audience to close their eyes and pray so he could get first dibs on the pickings. Rumor has it that the day after that incident the church fathers changed the seating in the church so the minister faced the back wall and the congregation faced the ocean. Of course, I have heard variations on that story in almost every old community on the coast. At any rate, wrecking was a lively trade. By 1800, as many ships sailed around Monomoy as through the English Channel. Chathamites were so proficient at the ghoulish business that Rudyard Kipling memorialized them in his book *Captains Courageous*.

But Wreck Cove and the Powder Hole are ephemeral features. They were created when long fingers of sand were swept around Monomoy Point. The resultant sandbars formed great recurving hooks of sand that created and protected harbors like Wreck Cove and the Powder Hole. Eventually the hooks closed and the coves started to fill with rainwater to become freshwater ponds. You can see the remains of even older saltwater coves in the freshwater ponds that presently lie below Monomoy's old Life Saving Station.

So what did the good citizens of Whitewash do when an 1850s hurricane shoaled in the Powder Hole with sand? They did what any sensible nineteenth-century coastal dweller would do: They retreated. Undoubtedly, most of the inhabitants regretted losing their island homes, but they had no choice. They simply shuttered their primitive houses or barged them back to the mainland. Many of the buildings were eventually cannibalized by later Chatham families to build the summer shacks we so enjoyed as children.[1]

History came full circle when Monomoy was declared a national wildlife refuge in 1944. The shacks were removed as their owners died, and the refuge gradually purchased the entire island. Since then thousands of people each year continue to visit the island in their own boats or on cruises sponsored by the Massachusetts Audubon Society. Modern visitors can even stay in the Old Monomoy Light House Station to recreate the kind of experience we enjoyed as kids. As I write this piece, a new chapter is dawning on Monomoy Island. On Thanksgiving weekend 2006, Chatham's South Beach overlapped the island, so today Monomoy is connected to the mainland for the first time since a storm severed the connection almost fifty years ago![2]

I relate these memories not merely out of self-indulgence but to help explain my biases. I loved visiting that shack on Monomoy Island and know how much people will lose if they are forced to leave their homes on barrier islands. But the history of Monomoy illustrates the ephemeral nature of barrier beaches. Everything—wind, water, waves, dunes, beaches, harbors, and ponds—is constantly in a state of motion. Shoals become dunes, and saltwater coves become freshwater ponds. People must also learn to live with these changes, to enjoy the coasts when conditions are right, but to know when to leave when conditions change. Conditions have changed along the coasts, and the time to retreat is now!

Chapter 1

The Problem

Coastal Erosion, Global Warming, and Hurricanes

Coastal Erosion: Cape Cod

I learned most of what I know about coastal erosion as a kid growing up on Cape Cod. There were few rules in our family. You could get away with almost anything, but one rule was sacrosanct, "Never run down the bank in front of the house." Even as children we had a visceral understanding that this fragile remnant of the Ice Age was our last defense against the rising Atlantic.

There were other reminders. The first thing we would do every summer was to drive to the Outer Beach to see the changes wrought by winter storms. Nauset Beach might be fifty feet narrower, the dunes two feet lower.

During the blizzard of 1978 we saw entire parking lots swept into the ocean and watched "The Outermost House" drift out to sea. Henry Beston had written his classic natural history book by the same title in the lonely shack. It had been Cape Cod's best known literary landmark. We knew we were living on a caprice of geology, an ephemeral pile of sand that was slowly washing away.

In fact, Cape Cod presents an ideal experiment to study the effects of sea-level rise. Ten thousand years ago the sea level was four hundred feet lower, the world's temperature was five degrees cooler, and the coast was a hundred miles further east. The Wisconsin glacier had just dumped a pile of rubble upon this coast plain. Paleo-Indians were hunting caribou and woolly mammoth on an island the same size as Massachusetts but eighty miles further out to sea. The island had high bluffs, barrier beaches, conifer forests, and open heaths, much like those on modern-day Nantucket. But the climate was warming and the sea level was rising almost three feet every fifty years, six times faster than now. Today the remnants of that island are the shoals of Georges Bank, the fourth most productive fishing grounds in the world.

Nine thousand years ago the Atlantic washed over Georges Bank, exposing what we now call Cape Cod to the full brunt of the Atlantic. But Cape Cod then was very different from what we see today. There were no smooth barrier beaches, just an irregular pile of sand, gravel, and boulders as high as a hundred feet. Without the interference of the old island, waves from the southeast could pound directly against the rugged headlands of the outer cape. This initiated a period of rapid erosion that continues today.

The waves that crashed against the headlands also piled up water along the shore. The water had to go somewhere, so it formed longshore currents that flowed in the rips and runnels that paralleled the shore.

These longshore currents are the same currents that carry swimmers quietly down the beach on a hot summer day. The only time they become perceptible is when they encounter a break in the sandbars and turn directly out to sea. Then they become an ill-named but treacherous rip tide. I remember one July day on Nauset when thirty people were swept out to sea by an unexpected rip tide. Fortunately, most knew enough to swim to the side instead of directly into the powerful current, and the rest were rescued by lifeguards. The police cars, fire trucks, and ambulances that soon arrived were thankfully not needed.

But longshore currents also perform a more prosaic role. Working in collusion with waves, they move sand quietly along the shore. Each wave stirs up sand from the beach and suspends it in water, where it is carried parallel to the shore in the grip of the longshore current. The next wave redeposits the sediment back onto the beach as a delicate little filigree of grains but a few inches further down the beach. Each sand grain has made an elegant little loop as it moved downstream from where it was before. Even on the calmest summer day the beach flows silently past swimmers who remain unaware of the migration occurring in front of their blissfully naked feet.

The process makes Nauset Beach one of the most powerful sand transport systems on the planet. Every year the equivalent of a dozen football fields worth of sand piled eight feet high travels down this beach. The south end of Nauset grows two hundred to three hundred feet per year, or almost a mile every ten years. Astronauts sitting in the international space station can see the process from space. If we discovered something this dramatic on another planet, it would receive worldwide attention.

But the story of barrier beach formation does not end there. When waves drive sand onto the shore, the sand dries, and some of it is blown back onto the upper beach. There it accumulates behind tiny sprigs of beach grass to build sand dunes. Given enough sand and wind, the dunes can grow to be sixty feet high, and if they are not anchored down by vegetation they continue to migrate landward.

Sometimes portions of a barrier beach break off to form barrier islands or, if there is enough sand in the system, to create free-standing barrier beach islands in shallow waters. Today there are more than two hundred barrier beach islands that protect the Gulf and East Coasts of the United States. It is the most extensive system of barrier beaches on the planet.

In essence what happens along these coasts is that the sun's energy is translated into wind and wave energy that then pushes sand together to create sandbars, beaches, and dunes. These features are themselves molded by energy into slowly moving waveforms that slowly scour and flatten the coastal plain as they are nudged on by sea-level rise. The entire system acts like a giant broom inexorably sweeping the coast inland.

The process is more apparent south of Cape Cod, where the coastal plain is flatter. As you drive east of any of the major cities of the mid-Atlantic you encounter land as flat as Kansas. Soybeans, corn, and tobacco grow on land that once was scoured by previous incursions of the ocean. The only topography occurs when you reach the shore and see the beach and dunes have risen up in response to the wind and waves and are being nudged imperceptibly forward by the steadily rising oceans.

But the system has another trick up its sleeve. During severe storms the beach migrates landward by rolling over itself. Swimmers can see the results of this rollover in places like the South Beach of Martha's Vineyard. A narrow ridge of peat will appear in the surf after every major storm. The peat is the remains of the marsh that used to grow behind the barrier beach. The beach has rolled over the marsh like a tank track rolling over a battlefield.

I witnessed the results of such rollover in 2005. A northeaster had just uncovered a hundred-foot strip of peat on Nauset Beach. The tracks of wagon wheels crisscrossed its surface, and I could see the hoofprints of horses and oxen embedded within the compact soil. The tracks looked like they had been laid down yesterday, but how did they end up in the surf on the ocean side of the beach? Actually they had been laid down three hundred years before, when colonial farmers used horse-drawn wagons to cut and transport hay growing in the marshes behind the barrier beach. The beach had simply rolled over, migrating almost a thousand feet landward, all because of the two feet of sea-level rise that had occurred since the 1700s. Today the only evidence of those times is this transient strip of peat and the Hay Market subway stop in Boston where enterprising Cape Codders once sold salt-marsh hay to colonial city slickers.

But the most dramatic example of this continual process of rollover occurred in 1626 when the *Sparrowhawk*, a small barque carrying colonists to the Virginia colonies, was swept off course and wrecked in the quiet waters behind Nauset Beach. The Plymouth colony rescued its countrymen and Governor Bradford dutifully wrote up the disaster in his daily log. The incident was then forgotten until a winter storm unburied the bleached timbers of the *Sparrowhawk* in 1863. But the wreck was now on the ocean side of Nauset Beach. What had happened? The *Sparrowhawk* had remained in the same spot but during the intervening 237 years Nauset Beach had rolled over itself, first burying, then unearthing the sepulchral remains. The wreck was displayed on the Boston Commons for several months. Now it resides in the Plymouth Historical Society Museum in Plymouth, Massachusetts.

Despite such examples, it was only in the 1970s that coastal geologists finally proved that barrier beaches migrate by rolling over themselves in response to sea-level rise. It was an astonishing discovery.

More importantly, coastal geologists started to appreciate how much better barrier beaches were at protecting the mainland than artificial structures were. You can think of them as intelligent breakwaters. During a storm, waves pull sand off the beach and the beach re-forms into as many as three offshore sandbars. These sandbars can then dampen down the storm's energy far more efficiently than the original beach. The upper beach rolls over and the dunes are washed back, which also dissipates more energy.

After the storm passes, the beach, the dunes, and the sandbars all re-form but the entire system has moved landward by as much as twenty feet. In a matter of months the system is able to protect the mainland once again. As long as this system can bend, flex, and move, it will dissipate energy and re-form as an intelligent storm control barrier. Armor the beach so it can't move, and it will be as dumb as a wall—a seawall that will eventually be swept away.

Hurricanes: Too Much of a Good Thing

After Monomoy, hurricanes were my second greatest passion when I was growing up on Cape Cod. This was fortunate, because Massachusetts has a greater chance of getting hit by a major storm than any other state. This is because Cape Cod juts far out into the Atlantic, where it can get hit by both hurricanes in the summer and northeasters in winter. This tends to keep New Englanders honest. We remember our storms.

My first recollection of a hurricane is of the one that came in 1954. Cape Cod had not experienced a major hurricane since the great hurricane of 1938. Hurricane tracking was not very sophisticated then, so many people were not even aware that Hurricane Carol had accelerated unexpectedly after brushing past Cape Hatteras. She slammed into Cape Cod shortly after high tide on August 31.

My family was huddled in our house at the head of Pleasant Bay. We took turns standing at the window peering through binoculars. Our neighbors had spent the night camping on nearby Sampson's Island. It was too risky to try to rescue them by boat, but nobody had seen them all morning long.

Suddenly we heard the rumble of a heavy truck. It couldn't reach us; storm waves had surged over the roadway. We were sitting on an island. We donned rain slickers and ran outside to learn that the Army duck had been dispatched from the nearby military base to rescue our neighbors. We watched as the amphibious vehicle lurched down our driveway, then splashed into the roiling waters. We lost sight of the duck as it bucked and plunged in the mass of seething white caps. Half an hour later we spotted the duck returning, with our bedraggled neighbors wrapped in blankets, but still alive.

Hurricane Carol devastated much of Connecticut, Rhode Island, and Massachusetts. Her hundred-mile-an-hour winds blew the steeple off Boston's Old North Church and flooded Providence, Rhode Island, under twelve feet of water. Four thousand homes, thirty-five hundred automobiles, and more than three thousand boats were lost. Carol caused $6.4 billion worth of damage in today's dollars.[1]

We were still cleaning up after Hurricane Carol when Edna swooped down on us on September 11. This time, people were better prepared. Nobody tried to spend the night on any islands. Our lights went out again, but we had candles, and batteries for our shortwave radio. We listened intently to the broadcasts as the wind slashed the trees outside. Through the windows we could just see scraggly pines bent over sideways by the approaching storm.

Then something strange happened. Just as Edna passed over Nantucket, her eye split in two. The announcer reported that one of the eyes was over Chatham, traveling north up Pleasant Bay. We rushed to the window. Sure enough, we could just make out a small spot of sunlight almost lost amid the curtains of gray wind and blowing waves.[2]

Suddenly everything went silent. The eye had moved directly over our house. We raced outside to discover a beautiful summer day. The trees stood upright, dripping large crystalline drops of sunlit rain. The light had a warm yellowish glow and the air had an unnaturally hot, almost tropical feel. I seem to remember the smell of ozone caused by lightning, but perhaps I am mistaken. We spotted the tracker plane flying slowly against the clear blue sky. The Air Force had started tracking hurricanes shortly after World War II. The next advance would be weather satellites introduced in the late fifties.

We didn't have long to enjoy the spectacle. The far side of the eyewall was fast approaching. The trees, which had been blown over to the west, were being slapped back down sideways to the east. We were just able to scamper back inside before the wall of water enveloped us once again.

It is fascinating how your perspective changes with age. I spent most of my youth hoping for the next hurricane. When Hurricane Bob hit in 1991, I was married with two children and owned a house and car. There was no foolhardy racing outside to observe meteorological phenomena.

No significant hurricanes had hit Cape Cod since 1955. I had forgotten how to prepare for hurricanes in the intervening years. We spent most of our time doing everything wrong. I filled up the bathtub and buckets with water, forgetting that we no longer had a private well. Gravity from Woods Hole's water tower would ensure that we still had water even if the electricity went out. I spent hours putting tape on the windows. Nobody's windows were blown out by the Category 2 storm.

Bob struck in the afternoon, and we spent an hour in total darkness listening to huge oak trees crash down around the house. I heard one land on the car I had neglected to put in the garage. How was I supposed to know not to leave a car under an oak tree? We had grown up with supple pines!

The following day dawned warm and beautiful. The roads were covered with a carpet of bright green leaves the storm had stripped from the trees. The air held the geosmine odor of new-mown hay. Groups of people walked awestruck through the carnage. The only house that still had lights was owned by a fisherman who had the foresight to move his boat's generator into the basement.

A mini-tornado had cut a swath of destruction through the

woods beside our house. The separated fibers of the oak trees showed how the giant trees had been twisted apart as the tornadoes touched down and raced through. I would later discover that the twisted fibers made particularly good firewood. For several nights following the storm the stumps of the downed trees glowed with a sepulchral bioluminescence. The hurricane had exposed fox fire embedded within their rotting trunks.

Yellow jackets blamed humans for the destruction. Whenever you stepped outside they would make a beeline for your face. My daughter was stung three times. More people were admitted to the Falmouth Hospital than ever before. Almost all of the admittances were for yellow-jacket stings.

For the first few days after the storm, our neighborhood had the atmosphere of a summer camp. Nobody could use their computers; we saw pasty white Woods Hole scientists we hadn't seen in years. Everyone was outside cleaning up their yards and chopping wood. I salted away three years worth of prime oak firewood. But after ten days without hot food, electricity, or showers, the situation lost its charm. The leaves on the remaining trees hung brown and dead, and great piles of brush lined the streets. Trash had not been collected, so the smell of rotting food was everywhere.

But a few weeks after the storm I took a boat to Cuttyhunk Island and noticed new green leaves sprouting on the island's apple trees. This was curious for September. Later, forsythia and chestnut trees started to bloom on the mainland. It was almost autumn but it looked like spring. Nature was fighting back.

It is instructive to compare these three storms. The 1938 hurricane was a Category 3 storm that killed seven hundred people and caused $6 million worth of damage. Hurricane Carol was a Category 3 storm that killed one hundred people but caused $500 million worth of damage. Hurricane Bob was a Category 2 hurricane that only killed six people but caused $1.5 billion in damages. You can see from these figures that through better forecasting and measures like evacuations we have become better at saving lives. But hurricanes have become more costly because there are so many new houses on the coast. It is estimated that if the 1938 hurricane occurred today, it would cause $35 billion in damages, almost half the value of damages caused by Katrina in New Orleans.[3]

There are other interesting parallels. The 1954 and 1955 hurricane seasons were superficially similar to the 2004 and 2005 seasons. But why had there been so many intense hurricanes in the 1950s then virtually none until 1992? Were the storms of the 1950s just an artifact of my nostalgic memory, or had something really changed?

Actually, several things have changed. In the mid-1980s, William Gray, a hurricane researcher at Colorado State, discovered that there was a correlation between weather patterns in northern Africa and the frequency of hurricanes in the Atlantic. Gray showed that when the Sahara Desert is expanding and the Sahel region of Africa is hot and dry, few major storms strike the East Coast of the United States.

This pattern occurred from 1966 to 1992, when only three hurricanes greater than Category 3 struck the East Coast of the United States. From 1944 to 1960 when the Sahel was wet and stormy there had been fifteen major Atlantic hurricanes. They had peaked on Cape Cod in 1954 with hurricanes Carol, Edna, and Hazel and in 1955 with hurricanes Connie and Diane.[4]

Now researchers have discovered that there appears to be a further correlation with the El Niño–La Niña cycle in the Pacific and atmospheric and deep-sea current changes in the Atlantic. When the Pacific surface waters are cool they change the high-altitude jet stream that delivers more stormy weather to the Sahel region of Africa, creating hurricanes that strike the United States. We are presently in this era of more frequent hurricanes. It is expected to last for thirty-five to fifty years. This will make storms like Katrina more probable.[5]

But what about global warming? Except for a few errant academics under contract to retrograde energy companies, most scientists agree that the world is getting warmer. But even if you don't believe that global warming is caused by human activity, it is difficult to deny that temperatures have risen dramatically in the past fifty years. Global warming has particularly affected the oceans' surface water temperatures. As we saw with Katrina, surface water temperatures have to be higher than 80 degrees Fahrenheit, and the temperature must extend 200 feet deep to spawn hurricanes. Now we have these conditions for several months off Africa, where hurricanes are born, and in the Gulf of Mexico, where they can kick start a run-of-the-mill Category 2 hurricane into a devastating monster like Katrina.

In 2005, Kerry Emmanuel, an atmospheric scientist at MIT, showed that these higher ocean temperatures have led to an increase in more intense hurricanes. The 2005 season seemed to bear out both these discoveries. It included the most destructive hurricane on record, Katrina, and the strongest hurricane to ever hit the American coast, Wilma. It even included Catarina, the first hurricane ever reported south of the equator in the Atlantic Ocean. It also included the most storms in a single season, requiring meteorologists to run through both the Roman and the Greek alphabets to come up with enough names for all the hurricanes. Whether natural cycles or global warming is the prime force, it is clear that we have entered an era of more frequent and more powerful hurricanes. If you believe in self-limiting natural cycles, you conclude that it may last for thirty-five to fifty years. If you believe it is caused by global warming, you conclude that it may last far longer.

Of all the things scientists have discovered about global warming, this increase in the destructiveness of hurricanes is the most immediately alarming. Increasing carbon dioxide, melting ice caps, and sea-level rise might affect us in the future, but more powerful hurricanes can affect us next year.

It is storm surges that cause the worst hurricane damage. The primary cause of storm surge is the low barometric pressure at the center of a storm. This actually raises the water under a hurricane so it may be several feet higher than the surrounding ocean. This lens of raised water may be several miles in diameter. It is enhanced by winds and waves that are stronger on the right side of hurricane because they include the forward speed of the traveling storm. So people on land experience a storm surge as a sustained high tide with waves, rather than a single surge of water as the term implies.

It stands to reason that if we have more powerful storms in the future, they will also pack higher storm surges. Twenty-foot surges may be the norm rather than the exception. But this time, when hurricanes and storm surges come ashore, the sea level will also be higher. If the 1938 hurricane hit today, the ocean level would be about seven inches higher than in 1938. If Hurricane Carol hit today, the ocean would be about six inches higher than in 1954. Six inches might not seem like a lot, but it means that storm surges will flood several miles further inland and barrier islands will roll over more rapidly and migrate further toward the mainland. If the Greenland cap collapses precipitously, as many experts fear, sea level and storm surges will be several feet higher in just a few decades.

Add to all this the fact that there has been a building boom along the coasts, and you have the ingredients for the disaster that devastated Florida and the Gulf of Mexico in 2004 and 2005. As the old ecologist of the Okeefenokee swamp, Pogo, once intoned, "We have met the enemy and he is us." With these wise words in mind, let's proceed to the Gulf of Mexico to learn how to live with this new reality.

Chapter 2

New Orleans

The Big One: 2005

On August 23, 2005, at 5 P.M., the National Hurricane Center in Miami announced the formation of tropical depression 12. Max Mayfield rubbed his eyes and turned away from his computer screen. The director of the center had a second sense about these things. He took another swig of coffee, adjusted his glasses, and scrolled back through the development log.[1]

The system had started as a wave of easterly trade winds wafting lazily off a stormier-than-usual North African coast. The easterlies had blown steadily over a warmer-than-usual Atlantic, whose surface waters were over 80 degrees, with her thermocline over 200 feet deep. Max was worried about the thermocline. It is the silvery interface that separates the warm surface waters above from the cooler waters below. Sometimes you can actually see it when diving underwater. This thermocline was so deep that storms could not churn up the waters to dissipate their heat. Now the surface waters were hot and deep. They could fuel potential storms for several more months.

As the easterlies converged under the influence of the earth's

THE NEW ORLEANS AREA

rotation, they created a tropical disturbance that had spawned a string of thunderstorms along its eastern edge. Warm tropical air inside this nest of swirling storms had created a column of low pressure that pulled the thunderstorms in closer. As the air inside the column condensed into water droplets, it released its latent heat energy, which lifted more air into the upper atmosphere ten thousand feet overhead.

Over the Bahamas the process had become a chain reaction, constantly lowering the air pressure inside the depression and producing massive clouds above it. The system had become a giant heat pump swirling counterclockwise under the influence of the earth's rotation. This was called the Coriolis effect.

On August 24, Mayfield had to upgrade the depression into tropical storm Katrina, and its winds were now circling her core.

The fourth tropical storm of the season was developing its own distinct personality.

By August 25, Katrina was a full-fledged but still minor hurricane. Clouds billowed over its center and bands of rain spiraled out crazily overhead. An eye was developing inside the core. The temperature inside this baleful center was an uncomfortable 80 plus degrees, with the air in the adjacent clouds close to freezing. The difference between these two temperatures provided the energy to draw the storm in tighter, increasing the speed of the surrounding winds like a figure skater pulling in her arms to spin yet faster.

Throughout the night of August 25, Katrina killed fourteen people as she swept across Florida before exiting the Everglades, once again as a downgraded tropical storm. But this was as expected. What really bothered Mayfield was a mass of deep hot water waiting for Katrina in the Gulf of Mexico. Usually this layer of superheated water was only a few meters deep. Each storm would break the layer up as it passed over the Gulf. But this mass of water had looped off the Caribbean Current, then sat in the Gulf growing thicker and fatter as it basked in the ninety-degree heat for several months. Now it was well over two hundred feet deep, and no storm could break it apart. But Katrina could feed off this bottomless supply of heat until she was several hundred times more powerful than a thousand well-placed nuclear weapons. Mayfield picked up the phone and dialed Walter Maestri, head of the Jefferson Country emergency preparedness office in New Orleans.

"Walter, this is Max. I'll make this quick. You have a dire emergency on your hands."[2]

That was all he had to say. Walter knew that Mayfield was a low-key kind of guy. If Max said it was a dire emergency, it was probably pretty near catastrophic.

Fat drops of rain were already starting to pelt the peaty soil of Yscloskey, Louisiana. Joe Gonzales took one last look at his house and the unfinished oyster boat he had just lashed to the ground with four thick hawsers. He had laid every piece of pecan flooring in his house and planked every piece of cypress on his boat. His only wish was to live long enough to take his bateau out onto the bayou one last time before he died. But today was not the day. He swatted at a cloud of gnats.

"Maybe the storm will get rid of some of these damn flies."

His wife said nothing and got into the car.[3]

On August 27 at 9 A.M. officials in Plaquemines Parish called for a mandatory evacuation of all citizens in the area. Some could still remember the floods of 1927, when a group of wealthy New Orleans bankers had convinced Calvin Coolidge to dynamite Plaquemines' levees in order to protect the financial welfare of New Orleans. The crevasse had left ten thousand Plaquemines families homeless. But each family had received $169 in compensation for their loss!

On August 27 at 2 P.M., wealthy white families on Dauphin Island in Alabama shuttered their summer homes and drove inland.

On August 28 at 4 A.M., Major Greg McHenry was trying to fly his WCJ-130 Turboprop into the eye of Katrina. But a sudden downdraft almost lurched the wheel out of his hands, sending the light plane careening toward the Gulf of Mexico. Lunch bags and equipment sacks flew around the cabin and Greg's small crew of hurricane chasers were momentarily weightless or pinned against the cabin ceiling. McHenry was finally able to gain control of the plane and wrestled it on toward the eyewall looming ahead.

"Dropsond away."

In the rear of the plane, Sgt. Jay Latham ejected the miniature weather station through a chute in the hull. All eyes scrutinized the computers as the dropsond parachuted down through the maelstrom.

"Wind speed 191. Wind speed 200."

At 4:21 A.M. McHenry pushed through the eyewall.

"Holy shit! The pressure is 915 milliners."

"No way."

"Three hours ago this baby was 935 milliners in the eye, now it's 915. She isn't losing strength, she's gaining it!"[4]

By the time the crew made their fourth pass through the storm the sun had risen. McHenry looked around him. They were flying through a Category 5 hurricane. Angry gray clouds castellated thousands of feet above them. This was what meteorologists call the stadium effect. It was certainly impressive, but what McHenry liked the most was the forty-mile disk of blue sky hovering above the swirling cylinder of clouds. The fourth strongest hurricane to ever strike the American coast was now aimed at America's most vulnerable city.

Mayfield made sure the National Weather Service's bulletin was to the point and blunt.

"Katrina is now a potentially catastrophic Category 5 hurricane. Preparations to protect life and property should be rushed to completion."

On August 28 at 10 A.M. Mayor Ray Nagin opened his unusual Sunday morning press conference.

"We're facing the storm that all of us feared. . . . I am hereby ordering everyone to evacuate the city."[5]

The day before, the mayor had merely requested a voluntary evacuation and announced the opening of the Superdome as a shelter of last resort. He had advised people to "Plan as if you're going on a camping trip, bring your own food and drinks and other comforts such as folding chairs. But no weapons and no large items, just bring small quantities of food, enough to last for 3 or 4 days to be safe."[6]

On August 28, around 5 P.M., Marty Montgomery ordered a drink inside Johnnie White's bar. He had done everything he could think of to prepare for the storm. Electricity was going to go out anyway, so he might as well get drunk. The bar was his second living room. It had never closed for sixteen years. It had stayed open for Camille and Ivan; why should it close for Katrina?

"So what are you gonna do if looters come, Marty?"

"Oh, I got ole 'kindness' up in my closet. She's the 12 gauge my Daddy gave me. If anyone tries to break in, I'll just say I killed 'em with kindness."[7]

Katrina struck at 6:10 A.M. on August 29. By evening the danger had passed.

People inside Johnnie White's bar were feeling blessed. A journalist dropped by to gauge the mood of the city.

"I told you we'd dodge another bullet. Tomorrow we'll start clearin' out the debris. You'll see, the whole city be cleaned up in time for the Labor Day parade."[8]

But Jeanne Meserve wasn't so sure. The electricity was out

and there was no external communication, but the CNN reporter had heard a rumor that water had mysteriously appeared in St. Bernard's Parish. It appeared to be more than the rain that had fallen during the storm.

An elderly woman in the Ninth Ward knew that the water was rising. It had flooded her out of her bed, then chased her up to the second floor. Now she was standing on a chair in the attic, using a board to keep her chin above the fetid water. Roaches kept running down the board and into her hair.

Something large and alive had brushed past her legs earlier in the night, and she had to hold very still when a rat swam by. She didn't want it to crawl onto her head to escape the water.

Just when Bernice felt she could hold out no longer, she realized the water had not risen in the past hour. In fact, it seemed to have gone down a little. She no longer had to crane her neck to keep her chin above water. Was the worst over?

Then it hit her. The tide had just gone out. She was surrounded with salty water from nearby Lake Pontchartrain, connected to the Gulf of Mexico. She was standing in her attic in New Orleans, six feet below sea level! She would remain trapped without food or water for three more days. Her neighbor ran down the street to rescue his brother's son. His body was found two weeks later, floating in the fetid waters. Others had seen water mocassins curled up in a ball on their beds. The whole world had witnessed the misery in New Orleans.[9]

On August 31, the price of oil soared to $70 a barrel and the price of gas jumped to $3 per gallon. The Associated Press carried a photograph of the remains of an oil rig beached on Dauphin Island. Behind it were the remains of 120 homes swept off their foundations. Their owners had spent a quarter of a billion dollars rebuilding their homes. The same houses had been destroyed and rebuilt four times in the past decade because of hurricanes. But it had always been easy to get money from the national flood insurance program, and the feds had always coughed up enough money to rebuild the bridge and road to the mainland. The same was true elsewhere: The same houses were hit and rebuilt after every storm. In nearby Louisiana, houses in Orleans and Jefferson parishes had accounted for 20 percent of the nation's total losses reimbursed by the federal flood insurance program in the past fifteen years.

By the end of the 2005 hurricane season, more than two thousand miles of coast from Texas to Florida looked like it had been carpet-bombed. A major American city had been destroyed and more than half a million people were being called environmental refugees. The term "environmental refugee" had been used before to describe mostly Third World people displaced from their homes by natural disasters. In the aftermath of the storm, many writers decried the use of the term because they felt it suggested that blacks displaced by Katrina were not full citizens of the United States. Other writers put the emphasis on the word "environmental" and defended the use of the term because it drove home the point that even in the United States natural disasters have the power to displace millions of people. The fact that sixteen months after the storm more than a quarter of a million people will probably never return to their homes in New Orleans makes me think that the term is more apt and certainly more powerful than the perhaps more accurate term "internally displaced citizens."

In the end, the 2005 hurricane season had been arguably more damaging than the Dust Bowl, the Chicago Fire, and the San Francisco Earthquake. The psyche of the United States had been rattled as never before. But would Katrina prove to be a watershed event? Would it actually change the way we live along the coast? Was it a precursor of more intense and more frequent storms? I decided to head to the Big Easy to find out.

New Orleans: The City Too Big for Its Breaches

I wasn't able to visit New Orleans until several months after Katrina, and due to the vagaries of modern aviation's "hub-and-spoke" travel, I had to fly to Detroit before proceeding due south to Louisiana. But the detour had its advantages. Below me the Mississippi River was busily transferring the sun's energy from the atmosphere, to the earth, and on to the ocean.

The early Mississippi started as a tiny trickle of meltwater cascading down a crevasse in an early Pleistocene glacier. It emerged from beneath the glacier, milky white with the powdery remains of crushed and pulverized granite. It drained the post–Wisconsin Age glacial Lake Agassiz, which was itself several times larger than all the modern Great Lakes combined.

For seven thousand years the Mississippi combined the litholitic soils of the North and the latsolic soils of the South, with the podzolic soils of suburban New York, and the black coal mine tailings of West Virginia. For seven thousand years it mixed the gray, brown, and red muds of thirty-one states and three Canadian provinces into the rich black topsoil of the Mississippi Delta and the fertile green marshes of coastal Louisiana.

For seven thousand years, the Mississippi braided back and

forth across floodplains, meandered through oxbows, and burst through its banks to discover new outlets to the Gulf of Mexico. For millions of years before that, a similar river flowed into a similar shallow sea. Its sediments buried billions of tons of rich planktonic ooze. Their accumulated weight cooked the microscopic creatures into sweet, rich crude petroleum, the black gold prize that powers modern civilization and has been the cause of far too many wars. Today the citizens of New Orleans pump this black essence of the sun's energy back up the Mississippi to fertilize the corn fields of Kansas and power the factories of Chicago.

But below me something has changed. The Mississippi can no longer jump its banks to discover new outlets to the sea. It can no longer deliver rich alluvial soil to delta plantations. It can no longer protect New Orleans with new marshes because the Mississippi is no longer a slow-moving natural stream meandering through a swampy delta. Today the Mississippi River is an artificially raised aqueduct sluicing unimpeded through two thousand miles of levees. The four hundred million tons of sediments that used to build up marshes every year now sluice straight over the edge of the continental shelf and into the dark abyss. The Mississippi no longer procreates lifesaving land; it spills its seed uselessly into the great maw of the Gulf of Mexico.

The infamous Lake Pontchartrain now looms up before us. From the air I can see that the "lake" is really an extension of the salty Gulf. We fly over the misnamed lake to approach New Orleans from the south. A broad expanse of trees and grass lies below us. But wait—the trees are all dead, and I can just make out the wake of an airboat below me. This land is actually marsh and swamp.

To the north is Lake Pontchartrain at sea level, to the east is Lake Borgne at sea level, and to the south is this swamp, only a few centimeters above sea level. Sitting in this leaky bowl twenty feet lower than the Gulf of Mexico lies the bustling metropolis of New Orleans. Winding through the Big Easy is the seemingly benign Big Muddy. But it is a charade, for here in New Orleans, the bulk of the Mississippi River is so vast that most of it glides by unseen 170 feet below sea level, while the rest of it looms twenty feet above the city, hemmed in by massive earthen levees. We know what happened when Katrina breached the levees protecting the city from Lake Pontchartrain. We can only guess what would happen if these river levees were breached and the full force of the Mississippi poured down into New Orleans.

No, this was not a good place to build a city. In the memorable words of the nature writer John McPhee, the spot was a place most campers would be loath to pitch a tent.[10] Yet not Jean Baptiste Le Moyne, Sieur de Bienville and special envoy to the king of France. Jean Baptiste had a vision. It was a vision of a city on a mighty river, a vision to rival that of Paris on the Seine.

Most thought Jean Baptiste seriously deranged.[11] But he ordered his engineers to design crude levees to control the Mississippi and reclaim the hopelessly flood-prone swamps so settlers could plant indigo and tobacco for overseas shipment.

The grandiloquent French and their later Spanish successors were soon replaced by Anglo-American plantation owners, who ordered slaves onto the riverbanks to muck up the rich black alluvial mud into three-foot, then four-foot, then six-foot levees. But after the British gave the new republic a fright in the Battle of New Orleans, President Madison dispatched the Army Corps of Engineers to help rebuild the city and fortify its levees in 1815. This would initiate the Second Battle for New Orleans, an epic struggle to determine how to control the Mississippi. The outcome of that battle would help determine coastal policy for the next three hundred years.

The objective of this battle was to open the mouth of the Mississippi. By 1855 the river had built up two large muddy sandbars across the mouth of the river, and trade was down to a trickle. On Congressional orders, the Army Corps of Engineers dispatched specially designed dredges to remove the natural blockades. But the thick, sticky sandbars were impervious to the dredges. In exasperation, the New Orleans Chamber of Commerce demanded that the federal government try a new approach. This time Congress turned to James Buchanan Eads, a self-taught civil engineer who was anathema to the Army Corps of Engineers.

As a young boy growing up on the St. Louis waterfront, Eads ran errands by day and read engineering books by night. From his research he realized that more cargo lay on the bottom of the Mississippi than on top of it, so he designed a fleet of salvage boats and started his own company. Business boomed. Steamboats continued to have a disturbing proclivity to blow up when their boilers failed, or to be sunk by running into "preacher snags," half-sunken old logs that bobbed up and down just out of sight in the swiftly flowing currents.[12]

Though he spent his nights studying engineering, Eads' real education came from the river itself. He designed a diving bell and spent hours walking the Mississippi's muddy bottom. He knew what it felt like to grope through billowing clouds of thick brown silt. He knew what it was like to be pushed and pulled by strong currents, never knowing when you might be sucked into a newly scoured death hole.

Eads used his river knowledge to design the St. Louis bridge, the world's first bridge made entirely from iron. It spanned the Mississippi at exactly the same spot where the young Eads had been introduced to the river, by being thrown into it unconscious when the steamboat he was on exploded as it approached the St. Louis wharf.[13]

The bridge made Eads famous. He was being compared to other great self-taught civil engineers like Leonardo da Vinci and Thomas Alva Edison, so when Congress called he was ready. He proposed letting the Mississippi do the work herself. He knew from his underwater walks how quickly the river could both build up and scour down sandbars. He designed two massive jetties that would jut far out into the Gulf of Mexico. They would channel the river's current into a powerful jet of water to scour a 350-foot channel through the muddy sandbars. At a depth of twenty-eight feet, the channel would allow even the largest oceangoing ships to steam upriver. He was nearly laughed off the waterfront for the outlandish scheme. But Eads was so sure of his carefully crafted calculations that he promised to pay for the project out of his own pocket. He guaranteed that he would only be reimbursed if the jetties worked. It was an offer no Congress could refuse.

Eads was opposed by Andrew Humphreys, the head of the Army Corps of Engineers, who had written what was considered the definitive study of the Mississippi River. But General Humphreys had also been appointed to his position because of his reputation as an effective but ruthless Civil War leader.

During a single charge he had lost 20 percent of his men in less than fifteen minutes. He wrote that the experience made him feel like a sixteen-year-old girl at her first party ball.[14] He was equally determined to crush the self-taught civil engineer who threatened to topple the comfortable monopoly on engineering that the Army Corps of Engineers had enjoyed since its incorporation under George Washington.

The dispute came to a head on May 12, 1876. General Humphreys had leaked information to the press that Corps of Engineers' own Major Howell had made soundings in Eads' channel, and it was only twelve feet deep. The news caused stock in Eads' company to tumble and made it impossible for him to obtain further loans.

Eads knew that his only chance for redemption lay in an oceangoing steamship that lay just offshore. It was the *Hudson*, a three-hundred-foot vessel that drew fourteen feet, seven inches. She was under the command of Captain E. V. Gagner, an old friend from their early days in St. Louis. Gagner welcomed Eads aboard, along with several journalists that Eads had invited to chronicle his gamble. They knew the stakes were as high as at any riverboat poker table back in New Orleans.

Gagner also knew how dangerous the situation was. The tide was falling fast and his local pilot had recommended that he not attempt to cross the bar. But Gagner did not hesitate.

"Head her for the jetties."

On shore, three hundred men ceased their labors to watch. They too knew the stakes, as the ship started to build up steam.

"Shall we run in slow?"

"No sir, let her go at full speed."[15]

Gagner knew that the increased speed would lift her bow a few inches above the surface of the water and push her stern a few inches below it. If Eads was correct they would just skim lightly over the mud, but if the Corps was correct the *Hudson* would tear out her hull and sink to the bottom.

The *Hudson* gained more speed, and a huge white wake billowed out ahead of her bows, then separated into a long "V" that sloshed to the edges of the willow-sided jetties. One of the journalists wrote, "As long as she carried that white bone in her teeth, the great wave that her proud bows pushed ahead of her as she sped forward—we knew that she had found more than Major Howell's 12 feet." [16]

Then she was through. Captain Gagner blew a powerful blast on the *Hudson*'s steam whistle, and three hundred men erupted in cheers that reverberated up and down the delta as the great ship made her way majestically upstream.

When Eads started his jetty project, less than seven thousand tons of cargo was being shipped from New Orleans to Europe annually. A year after he finished, that total rose to 450,000 tons a year, and New Orleans became the second largest port in the United States, trailing only New York in total tonnage. By 1995, New Orleans ranked first in the nation.

However, Eads' victory had several unintended consequences. The jetties were nothing more than levees that extended into the Gulf of Mexico. But they proved to be so effective that the Army Corps of Engineers adopted a "levees only" policy to control the entire Mississippi River. That meant no more spillways or outlets would be built to relieve pressure built up by floodwaters. Neither Eads nor Humphreys fully supported the "levees only" policy, but it became the Army Corps of Engineers guiding principle to both control the Mississippi and defend the American coasts with armored seawalls. It would also initiate the destruction of marshes that, had they been allowed to grow unimpeded for the last seventy-five years, would have dampened down the storm surges that destroyed New Orleans during Hurricane Katrina.

So, from the very beginning, levees have been both New Orleans' solution and her nemesis. Every time the levees are raised, people build new homes. Every time the levees are breached, the houses flood. But the levees and the Army Corps of Engineers have become indispensable to the continuing existence of the city.

With the levees in place, New Orleans grew larger, more important, and more dangerous to live in. Like every major city, New Orleans also had the political and economic power to alter the surrounding countryside to its own advantage. This affected the area's social climate. The early plantation owners became the old cotton families who went east to college, sat on the levee boards, and ruled the state and city through membership in the most prestigious krewes that ran Mardi Gras and in the city's exclusive Boston, Mystic, and Louisiana social clubs.[17]

It is said that the old cotton money still looks down on the new petroleum money, and that even the top students at the University of Tulane's Law School will not be made partners in New Orleans' best law firms unless they come from one of the original cotton families.[18]

Of course, the system bred resentment. Initially it was the resentment the poor, white, rural farmers felt against the citified whites who controlled the state. The situation had come to a head during the 1927 floods, when the president of the largest bank in New Orleans, who was the scion of one of the cotton families, coerced the governor to dynamite the rural levees of Plaquemines Parish in order to save downtown New Orleans. The city's banks were saved, but six thousand people lost their homes and were compensated the unseemly amount of $169 each for their suffering.

The sense of outrage was so strong that Huey Long, an unknown rural farmer, was able to unseat the incumbent governor. The "Kingfish" then ruled the state with an iron fist, bypassing the traditional New Orleans power brokers by levying taxes on the newly discovered oil and gas wells. He used this new revenue to construct schools and highways throughout the rural sections of the state. These public works projects made him the local hero to the poor, white Northern farmers and the "coonass Cajuns" who peopled the bayoued coasts.[19]

However, the Kingfish also made enemies in high places and was eventually assassinated on the steps of the state capital. But he had left his mark through his many improvements to the traditional laissez-faire political culture of patronage and corruption. If you were the governor of the state, you could get kickbacks from the sale of riverfront casinos. If you were appointed to the levee board, you and your family had a job for life. If you were elected to the board of assessors, you had the power to lower people's taxes in exchange for their vote and another New Orleans tradition, *un petit lagniappe*, a small gift given to preferred customers. While the schools and tax rolls might suffer, the corruption was shrugged off as merely another manifestation of the city's tolerance for flamboyant behavior.

However, the most flagrant example of the city's audaciousness lies three hundred miles north, where the Mississippi turns abruptly and unnaturally east. This is where Congress mandated that the Army Corps of Engineers stop time. The problem is that ever since the Ice Age, the Mississippi River has been able to writhe back and forth like a water moccasin swimming up a bayou. Every time the river writhed, it created a new outlet to the Gulf of Mexico and built up a new delta. Today, most of the southern coast of Louisiana is composed of the remains of these former deltas.[20]

Ever since the 1927 floods scoured out a new channel, the Mississippi has wanted to writhe back south as it has every thousand years or so. The river has moved as far east as it can possibly go; any further and it will have to start flowing uphill. Now the Mississippi wants to join the much younger Atchafalaya River that surges impetuously south, into the Gulf of Mexico.

Most geologists say it is simply a matter of time before the rule of gravity takes over and the Mississippi flows into the waiting arms of the Atchafalaya. But Congress has a way of ignoring laws of nature to create its own reality. In 1954 it passed a law that required that no more than 30 percent of the Mississippi should ever flow into the Atchafalaya. This was the way it was in 1950 and this is the way it should always be. In effect, Congress had mandated that time stand still.

Congress ordered the Army Corps of Engineers to build an elaborate system of locks and gates to ensure that the Mississippi's obsession never achieves fruition. Today the equivalent of seven Niagara Falls plunges over the dam from the Mississippi into the Atchafalaya, making her America's second largest river by volume and the seventh strongest and most treacherous river in the world. A fortified tugboat patrols the area twenty-four hours a day to prevent two-hundred-foot barges from being sucked into the maelstrom below.

Geologists contend that the Army Corps of Engineers is fighting a losing battle. During the 1973 floods, engineers opened all the gates and watched in horror as roaring waters washed a massive guide dam down the river and the main dam vibrated so violently they feared it would collapse. Vibrations during an earlier flood had caused coal to ignite in a nearby railroad car. After the flood passed, the Army Corps of Engineers discovered that the torrent had scoured a hundred-foot hole in front of the dam. If the Mississippi ever does shift its main channel into the Atchafalaya, an arm of the Gulf of Mexico will transgress north to inundate both Baton Rouge and New Orleans.[21]

So there it is: the brief history of a city too important to abandon, yet a city whose continued existence depends on solutions that only make matters worse. It is the history of what engineers call an irreversible mistake; it is the story of a city that has grown too big for its breaches.

First Decisions

Immediately after Hurricane Katrina hit, it became clear that New Orleans had to deal with two interlocking issues—levees and rebuilding. Officials handled the first by jawboning for a $32 billion federal handout to protect the city against a Category 5 hurricane.[22] But it was clear that the rest of the United States was not going to shell out that kind of money for levees that had already failed. On October 28, 2005, the Administration announced it would spend just $1.6 billion to rebuild the levees back to what they had been before Katrina. This put the rest of the country on notice that the federal government was not prepared to protect every vulnerable city against a Category 4 or Category 5 hurricane. It also guaranteed that homes rebuilt in flooded areas in New Orleans could be destroyed again in the face of future storms. There the issue rested for several months.

The second problem involved deciding which areas of the city should be rebuilt. Mayor Nagin used a time-honored method to avoid making such decisions. He appointed a blue ribbon panel. The Bring Back New Orleans Commission immediately went to work, hiring a nationally recognized think tank to come up with the preliminary plan. The Urban Institute's proposal was a well-thought-out plan that called for rebuilding on the oldest highest ground first, then gradually working back down to the more vulnerable lower lying areas. Eventually the lowest-lying, most vulnerable areas would be turned into parks and wetlands to soak up water in the event of future storms. The plan essentially recreated the early history of building on the city's natural high ground first. The plan was dead on arrival.[23]

Hundreds of people turned out to shout down the proposals, and angry mobs blocked demolition vehicles from removing the remains of condemned buildings. Cynthia Willard Lewis, a representative from East New Orleans, told reporters that her neighbors were not going to be shoved to the back of the bus. In response, the City Council passed a resolution that all residents would be able to return and rebuild their homes in exactly the places they had formerly been. Mayor Nagin was nowhere to be

seen. He was on vacation in Jamaica, though his office had announced he had to be away on business in Washington.[24]

The commission had also been talking to Wallace, Roberts,
and Todd, another planning firm from Philadelphia, which dis-
missed the Urban Institute's plan as "planning for failure." The
firm argued that the city's future footprint should be determined by the New Orleanians themselves and recommended
that the city take a four-month moratorium on construction so
that neighborhoods could hold meetings to prove that enough
residents would return to warrant the rebuilding of infrastructure. Nobody bothered to ask how such meetings would be held
while most of the residents were still scraping by in Houston and
Atlanta.[25]

The reaction to this second plan was equally explosive. This
time, Mayor Nagin supported the proposal for eight days, then
withdrew to announce that he was a property rights man and
everyone would be invited back to New Orleans to rebuild. The
market would ultimately decide which neighborhoods would recover and which would be abandoned. The mayor was sounding
so much like a Republican that his opponents started to refer to
him by his old nickname, Mayor Reagan Nagin.

The result was instant chaos. It was everybody for himself.
Building permits were handed out like candy and the city became a jack-o'-lantern pattern of a few individual homes being
renovated amid a darkened background of rubble and abandoned
buildings. Thousands of well-intentioned volunteers pitched in
to rebuild vulnerable homes in exactly the spots where they
could be flooded again.

Sean Reilly knew there had to be another way. The former

state legislator had given up on politics to return to Lamar Advertising, the nation's third largest billboard firm, which his family had started. But when Governor Blanco asked him to join the Louisiana Recovery Authority he was only too eager to jump back into the fray. The Louisiana Recovery Authority was the state's counterpart to the Bring Back New Orleans Commission, and would prove to be one of the most effective of the many post-Katrina recovery organizations.[26]

After each meeting of the Bring Back New Orleans Commission, Reilly would patiently explain to the press that the state would not let the city put people back into areas where they could be flooded again. But from his work in the state legislature, Reilly knew that the real problem was how the Bring Back New Orleans Commission was run. Too many people used it as a forum for grandstanding. What you really needed was a quiet room where a few people could hash out their differences out of the limelight, then present their plan as a united front.

Reilly shared his concerns with his friend Walter Isaacson, the vice-chairman of the Louisiana Recovery Authority. Isaacson was born and bred in the Broadmoor section of New Orleans. In his youth, he whiled away his afternoons discussing books with Walker Percy on the porch of the esteemed Southern writer's home overlooking the magnolia-lined Bogue Falaya River. During the early 1970s Isaacson studied government at Harvard during his winters and wrote for the *Picayune Times* during his summers, while rereading his Tennessee Williams and William Faulkner.[27]

Time magazine hired Isaacson as soon as he finished his Rhodes scholarship at Oxford, and the gifted young writer soon ascended through the ranks to head the magazine in 1995. In 2001 Isaacson was hired as the chief executive of CNN to complete the journalistic hat trick. In the interim he had also managed to author books on the Founding Fathers, Henry Kissinger, and Benjamin Franklin. But he never forgot his roots, so when Governor Blanco invited Isaacson to serve on the Louisiana Recovery Authority, he readily accepted.

Isaacson's most important meeting occurred the day after the Administration shot down Louisiana Congressman Richard Barker's proposal to create a national building authority to buy back flooded homes and sell them to builders for resale.

Isaacson stormed into the White House to meet with the equally tempestuous Karl Rove. It was a meeting of polar opposites. Rove was already ticked off at Isaacson for his repeated charges that the White House disdained New Orleans because it was a predominantly black city. Accusations flew, tempers rose, and several hundred cubic meters of high-octane oxygen were sucked out of Rove's West Wing office.

Eventually Rove ended the meeting with the suggestion, "Why don't you just come up with a simpler plan, something direct to homeowners?"[28]

Isaacson fumed the whole way back to his office at the Aspen Institute in Washington. But several months later he was able to laugh about the meeting with a reporter from *Fortune* magazine. "You know what? In retrospect they were right in the White House. And we were wrong. It was too complicated."[29]

Isaacson had come into the process favoring a top-down planning solution. But after his family's home in Broadmoor was slated for demolition, he started to think that the correct role for planning officials might be to help neighborhoods decide how to rebuild. It was not just a matter of whose ox was being gored. The Louisiana Recovery Authority had seen that the top-down approach did not work. It was looking for a new approach, and if Republicans were the ones making the decisions, it might as well back a Republican proposal.

In January 2006 it was Reilly's turn. He flew to Amarillo, Texas, to meet with Donald E. Powell, the President's new appointment to the federal recovery effort. Louisianians had initially been skeptical of Powell. The wealthy Texan had been a close personal friend of the first President Bush and top fund raiser for the second. After their experience with the former head of FEMA (the Federal Emergency Management Agency), Michael Brown, New Orleanians had been understandably skeptical of the President's political appointments.

But folks in East Texas could have told them differently. They had known Donald Powell as a hardworking young kid from a blue-collar neighborhood of Armarillo. As a boy he sold chewing tobacco throughout the Panhandle, then won a football scholarship and went on to become a Texas banker—a special breed of banker known for their straight-talking, no-nonsense style.

That style initially landed Powell in a lot of trouble. In one of his first meetings, in Baton Rouge, the new appointee was asked about the federal government's responsibility to protect people living on floodplains. The wealthy businessman responded with his usual East Texas twang, "Well, when I built my house outside Amarillo, I was responsible enough to buy my own flood insurance." [30] Even for an audience of polite bankers, that was too much to take. Boos erupted. The President had sent them another political crony. Didn't this guy know the difference between the bone-dry Panhandle of Texas and the sinking coastline of Louisiana?

Shortly after the well-publicized debacle, Powell set out on his own Texas-style walkabout. He wandered through the devastated streets of New Orleans in worn jeans and old cowboy boots, simply talking with people about their problems. He listened to the maid who made his bed at the Sheraton Hotel and called Senator Ted Kennedy's wife Victoria Reggie after he learned she hailed from Louisiana. He joined building inspectors looking at high-water marks in the Ninth Ward and became so well versed in people's problems that when President Bush visited the Ninth Ward in March, Powell was able to point to specific homes and reel off the woes of their distraught owners.

Powell and Reilly had their lunch in Amarillo at a local eatery with tables bedecked with thin, white, paper tablecloths. Reilly made the case that if you lived behind a federally warranted levee you should not be penalized for not having flood insurance. In fact, the federal flood insurance program didn't even provide insurance to people behind levees because they were considered to be protected from floods. If the levees had failed, the government should be responsible for reimbursing homeowners because the government had effectively promised homeowners that they didn't need flood insurance. Gradually Powell saw the light.

By the end of lunch their tablecloth was covered with a maze of numbers and scribblings, detailing house-by-house data on the losses suffered in each neighborhood. The final result would not be announced until February, but the process had started with two people meeting out of the glare of the media to hash out their differences. In the end it would mean that New Orleans would have an additional $10.4 billion to rebuild homes and businesses through the Louisiana Recovery Authority's Road Back program.[31]

"All the King's Horses and All the King's Men":
Donald Powell, Spring 2006

On March 30, 2006, Donald Powell told Louisiana that as far as he was concerned New Orleans had no levees. Even after the Army Corps of Engineers finished spending the $3.4 billion earmarked to rebuild the levees, the levees would still allow waves to overwash their tops. Therefore, the levees could not be considered to provide any protection from a hundred-year storm. That would require an additional $6 billion.[32]

The problem was that Katrina had changed the definition of a hundred-year storm. The process had started right after the 2005 hurricane season when a major modeling firm held a special meeting in Bermuda. At that meeting, Risk Management Solutions made the dramatic decision to scrap the old way of modeling hurricanes based on historical data to one based on recent evidence of climate change. The old method used 106 years worth of hurricane data compiled in 1925. The new method was based on data starting in 1995, the year when scientists believe the world definitively entered the new pattern of more frequent and more powerful storms.[33]

The insurance companies grasped at the new method like a drowning man lunging for a life preserver. The new method meant they could justify raising their rates to offset catastrophic losses expected in future hurricane seasons. While politicians and scientists could afford to make sure that every "i" was dotted and every "t" crossed, the insurance industry didn't have such luxuries. They had to be able to change policy quickly in the face of new realities. The decision reverberated through numerous government bureaucracies, among them the Army Corps of Engineers.

Since the 1920s, the Army Corps of Engineers had based their construction of New Orleans' levees on the old model. In essence they were rebuilding levees based on a hundred years of data compiled in 1925. In addition, most engineers designed their projects to have a fudge factor; for instance, offshore drilling platforms are designed to withstand waves that are four to six times larger than the theoretical maximum. But the New Orleans levees only had a safety factor of 1.3. This fudge factor was based on criteria devised in the 1940s to protect agricultural land from flooding. By applying the same standard to the city's levees, the Army Corps was effectively treating the people of New Orleans as if they were no more valuable than cows![34]

Donald Powell argued that if the Army Corps of Engineers were given enough money to rebuild the existing levees based on new methods and new safety factors, he could accept the hypothesis that they would not have breached during Katrina, and that New Orleans would have only been flooded under the three feet of water that slopped over their tops.

Powell's complex argument delighted local politicians. It seemed that someone in the federal government was finally getting it. Perhaps New Orleans had a new hero willing to speak the truth and stand up to their old nemesis, the Army Corps of Engineers.

Powell told a reporter for the *New York Times* that he arrived in New Orleans with no preconceived thoughts, but that he gradually realized that while Katrina was an act of God, what happened in New Orleans was an act of both God and man. The levees breached because there had been flaws in their design and construction. In June the Army Corps admitted as much in a six-volume, six-hundred-page mea culpa that became an important part of the healing process for New Orleanians. The interagency performance evaluation also suggested that some parts of the coastal system were so damaged and so subject to future impacts that they should not be engineered to support future habitation. It was a sign that things were starting to change within the old water dinosaur of a bureaucracy. The report would have significant impact on future decisions up and down the coasts of America.

Powell made good on his conversion by convincing the President to have Congress appropriate an additional $4.2 billion to rebuild New Orleans' levees. However, now he admitted that even that was not enough. At least $6 billion would have to be spent to rebuild the levees before he and FEMA could even consider they existed.

Six billion dollars was still a far cry from the $32 billion New Orleans had initially wanted Congress to appropriate to construct levees and barriers high enough to protect against a Category 5 hurricane. But by now, everyone knew that was simply not going to happen. The United States had made the tacit decision that no single nation could afford to fight all the effects of global warming. You couldn't just build seawalls to protect every coastal city against a Category 5 hurricane; Category 3 is about as high as any nation could afford. Beyond that, affected areas would just have to evacuate and hope for the best. That decision has become the tacit policy of the United States. Most nations cannot even consider building defenses sufficient to protect their coastal cities against even a Category 3 hurricane.

However, all in all, things appeared to be getting better in New Orleans. One in ten businesses had reopened and every day more people returned than were expected to. But among them were some of the drug traffickers who had caused both the white and black middle classes to flee the inner city. Before the storm, these violent criminals had been accepted as just another part of this complex city. Now neighbors were more willing to drop a dime if they saw a suspicious person entering an abandoned building. Nobody wanted the dealers to re-establish a toehold in neighborhoods that had been without crime since Katrina.[35] The same was true for the schools. After years of letting schools slide, parents had a new determination to make them work. While *laissez les bons temps rouller* was a fitting theme for the city without a care, there was a growing realization that it was not the way to run a decent school system.

And of course, the city was gearing up for another election. To outsiders, New Orleans might be known as the city that knows how to throw a party, but insiders knew that what New Orleans is really good at is holding an election.

The 2006 mayoral election was going to be crucial. Katrina had forced most black people to leave the city. For the first time in several decades, there were more whites in New Orleans than blacks. Activists called it the "Surge and Purge" school of urban planning. However, if blacks really wanted to hold the mayor's office they would have to vote for the conservative Mayor Nagin. Many whites feared that Nagin's ultracasual style would drive away future federal aid.[36]

But the greatest problem the city faced was still uncertainty. Powell had signaled that when the FEMA maps finally came out, they would be the definitive planning tools, determining whether someone could get flood insurance, whether they had to rebuild their home fifteen feet higher, or whether they were eligible for the $150,000 rebuilding grants.

New Orleanians expected the worst, but they were in for a big surprise. On April 12, 2006, Donald Powell announced the long-awaited flood rebuilding advisories. Everyone assumed the advisories would recommend that people rebuild their houses at least eight feet higher than before. Some federal officials had even been hinting at figures as high as thirty feet. Instead, the advisories stated that people would only have to rebuild their houses three feet above ground level. That could still add $60,000 to the cost of rebuilding a home, but it was a pittance compared to what it would cost to raise a house thirty feet higher.

But did the advisories make any sense? FEMA tried to make the argument that they were the result of a complex formula that incorporated the Army Corps of Engineers levee performance analyses, historic rainfall data, and computer models of storm surges, But what the advisories really represented was a woeful lack of responsibility. Nobody had been able to make the hard decisions about which neighborhoods should be rebuilt and which should revert to nature to protect against future floods.

Instead, everyone would be allowed to rebuild the same death traps they had lived in before.

There comes a point when government has to step in and protect people from themselves. States do this when they protect citizens from driving drunk or borrowing money at subprime or exorbitant rates. This time, the federal government failed. The new improved levees that Powell had successfully lobbied for would not be complete until 2010. That would leave the city entirely vulnerable for four years. After that, homes that had flooded under thirty feet of water were expected to be protected by being rebuilt three feet above grade!

What had happened? Neither FEMA nor the Army Corps of Engineers retained enough credibility after Katrina to be able to make strong responsible decisions. They had seen what happened to the city's plan for rebuilding, which had run into a political firestorm after it named specific areas to be abandoned. Such plans could so easily be labeled racist that the federal government had decided that it had to devise a policy that appeared to affect everyone equally. Instead, it put the same most vulnerable people back in the same most vulnerable situations. The policymakers forfeited what their managers needed most: good, hard, objective data on which to make difficult but responsible decisions. Instead, they had agreed on a compromise based on a convenient fiction, the myth that Katrina had been a one-time event, the belief that if the levees had only been built better the city would have never been flooded, the hope that only three feet of water will spill over the top of the newly built levees in the event of another Katrina.

The feds fell into the same trap that they fell into with the "levees only" policy adopted after the 1927 floods. Then, nobody had been able to make the hard choices so they took the expedient path. The same was true in 2006. None of the federal agencies really wanted this policy, but it was the only one they could all agree on.

The advisories made it clear that the federal government hadn't really learned anything from Katrina and that people were going to be left just as vulnerable to major storms as before. The suspicion lingered that Donald Powell was the architect behind this well-meaning but ultimately risky policy change.

Shrimpin' the Full Moon Tide

It is May 13, 2006. There is no wind, there are no sounds, save for the quiet droning of a shrimper motoring down Barataria Bay. A string of pelicans glide on motionless wings against a pastel sunset.

The shrimper has been watching the moon for the past few days. Tonight the moon will be working in collusion with the sun to draw the spring tide out through the quarter-mile pass between Grand Isle and Grand Terre.

The shrimper has motored past row upon row of dying oak trees and strings of telephone poles standing in two feet of water. Only a few months before, they had been on dry land. This was not because of Katrina, just the normal sinking of land from lack of sediments, and 10,000 miles of canals dug in the marsh to lay oil and gas pipelines. The channels start out only 35 feet wide, but in less than a decade they erode to more than 200 feet across. Then they allow salt water to sluice in and out of the estuary easily, undermining peat and tattering the marsh. Everyone can remember seeing portions of their back lots slowly slumping into the ever-expanding bay.

This entire coast loses 25 square miles of marsh every year, the equivalent of losing an entire football field every three minutes. That was before Katrina tore out an additional 118 square miles, which was only a trifle compared to the 2,000 square miles of marshlands lost since the levees were built after the 1927 floods. That is the equivalent of losing a state the size of Delaware in less than eighty years.

Now the shrimper sees other boats converging on the gap between the islands. They stagger out from as far away as Lafitte, Duloc, and Leeville, all hit hard but recovering from Katrina. The pass is already cheek by jowl with shrimpers. Giant booms hold their nets outstretched. The sound of the engines is deafening. The shrimpers have to keep their engines running at full throttle just to hold themselves steady against the outgoing tide. They have lined up in a "V" formation facing into the onrushing water. The boats are so concentrated that they can be easily seen on Google Earth.

The shrimper breaks through the line and comes up from the rear to jostle for position. Men sporting wrap-around sunglasses and colorful bandanas give him a wide berth. He has been known to knock competing fishermen off their feet with a single throw of a well-aimed shackle. The great-great-grandfathers of some of these same fishermen probably joined Jean Lafitte's pirates during the Battle of New Orleans.

But tonight their prey are brown shrimp, *Farfante paneus aztecus*, that have been waiting for darkness to ride the tide back out into the Gulf of Mexico. Since April they have grown as much as an inch every week in the LaFourche estuary. Now they are ready to return to the Gulf to spawn and lay their eggs.[37]

Overhead a fight breaks out. The shrimper has been trying to maintain his position in one of the rips where the current is strongest and the fishing is best. But one of the younger captains maneuvers too close and has to be firmly rammed to be convinced this is not his territory. Water hoses are aimed, but there is no time for retaliation. The moon is now high and boats on all sides are hauling in hundreds of pounds of snapping, popping, spiny-skinned shrimp.

At about three in the morning the tide turns and scores of boats have to zigzag and scramble to avoid being swept back into the estuary, where the shrimping season has yet to open. A few more fights break out, but the fishing continues. A neighboring boat is famous for having stayed out all night in a hurricane so the crew could continue fishing; a minor Cajun cluster fest is not going to deter them.

Just before dawn, the shrimp stop moving and the shrimpers steam back toward Leeville. It has been a good night. Many of the boats caught and cleaned close to 2,000 pounds of shrimp. It has been that way ever since the levees were built in the 1930s. Every year the marsh eroded it meant twenty-five more square miles of detritus, twenty-five more square miles of bacteria, and twenty-five more square miles of fecal pellets: all abundant food for the record catches of brown shrimp, white shrimp, pink shrimp, oysters, and blue crabs.

This year the catch is particularly good because Katrina left that much more dead offal in the water. She also destroyed so many boats that there are more fish to go around. It has often been said that world wars and hurricanes make the best fisheries management tools: Remove fishing pressure off a fast-growing fecund little species like shrimp, and they will rebound to historic heights in a single season.

But how long can this artificial bonanza continue? In a matter of decades, Louisiana will run out of marsh and the extended boom will come to a crashing end. A major food source will be lost and a way of life will cease to exist.

However, there is a plan that may both solve the many interrelated problems facing the tattered Louisiana coast and provide long-term protection to New Orleans. It is the Third Delta Conveyance Channel, a project to divert one third of the Mississippi south of New Orleans to rebuild her marshes.

The project has much to recommend it. When it was originally proposed in 2000 it called for the Mississippi to do all the heavy work. The Army Corps of Engineers would simply divert the river at Donaldson, Louisiana, then dig two 200-foot-wide guidance channels down either side of the LaFourche Bayou. Within a few decades the river would erode the banks of the original guidance channels into two half-mile-wide rivers that would, in turn, deliver enough sediment into Terrebone and Barataria Bays to build two seventy-five-square-mile subdeltas. During the next fifty years these deltas would double in size to protect New Orleans, bring back the marshes, and eliminate the dead zone in the Gulf of Mexico caused by runoff fertilizer flowing down the Mississippi from the Midwest.

Since Katrina, planners have been proposing that the Army Corps of Engineers lend the Mississippi a hand by siphoning sediment off the bottom of the river and pumping it in pipelines directly and more quickly to the coast.

A smaller diversion project in nearby Caernarvon, Louisiana, has been building up twelve and a half square miles of new marsh every year since 1981. Since every mile of marsh dampens down storm surge by as much as a foot, in as little as five years the Third Delta Conveyance Channel could build up enough marsh to prevent the storm surge that flooded New Orleans in 2005. In twenty years the project has the potential to protect the city from storm surges generated by a Category 5 hurricane. That time frame is less than would be required to obtain the necessary permissions and build levees high enough to protect against a Category 5 hurricane. Plus, the Third Delta project would cost $15 billion less!

But there are always confounding problems with irreversible mistakes. The levees that the Army Corps of Engineers constructed after the 1927 floods not only channeled the Mississippi out into the depths of the Gulf of Mexico, but also reduced the amount of sediment carried in the river from bank erosion. Dams and soil conservation programs did the rest, effectively cutting off the flow of topsoil from the Midwest and coarse sand from the Rockies.

The bottom line may be that the Mississippi no longer carries enough sediment to rebuild the marshes faster than the land is subsiding and sea levels are rising. Together, the two processes cause the sea level along the Louisiana coast to rise more than two feet every fifty years. This may mean that the Third Delta Diversion Project will not work and that the three-hundred-mile-long Cajun Coast that supplies over a quarter of all the seafood caught in the United States will be lost in the next half century. But for now we must move on to Florida, which has its own problems with hurricanes and sea-level rise.

Chapter 3

Up the Coast

The Great Miami Hurricane, September 18, 1926

"Hurricanes are no more risky to life than venturing across a busy street,"[1] intoned Richard Gray after a minor storm clipped Florida's coast in July 1926. The head of the U.S. Weather Service station in Miami was just expressing the *zeitgeist* of his times. Technology had made natural disasters like hurricanes a thing of the past; what was important were human inventions, things like trains, automobiles, and electricity that were reshaping America. As a result, he paid little attention when an incipient tropical storm was reported a thousand miles east of Barbados. It was September 11, 1926.[2]

Hurricanes were the last thing on people's minds in the 1920s. The nation was in the fevered grip of Prohibition, and nowhere was the jazz age hotter than in southern Florida.

Glenn Curtiss, the inventor of the Curtiss Aircraft Engine, had sunk much of his considerable fortune into Hialeah, a 14,000-acre gangster paradise of illegal booze, illegal casinos, illegal jai alai, and illegal race tracks. Henry Flagler, John D. Rockefeller's right-hand man at Standard Oil, extended Florida's

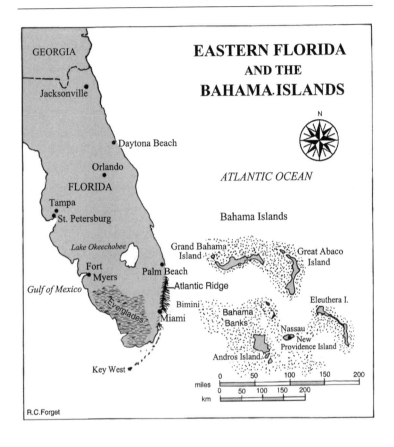

East Coast Railway from St. Augustine, to Palm Beach, and on into Miami, where he had donated land to Richard Gray's U.S. Weather Service so it would extol Florida's balmy weather to listeners huddled around their radios in the cold Northeast. Henry Ford was turning out a new Model T every ten seconds so newly wealthy workers could drive down the Dixie Highway on Florida's Atlantic Ridge. Businessmen, socialites, entertainers, and athletes took Flagler's trains south to stay at his flagship hotels:

the Ponce de Leon in St. Augustine, the Royal Ponciana in Palm Beach, and the Royal Palm in Miami.[3]

Things were a little slower on the Gulf Coast. In 1923 a group of motorists who called themselves the Tamiami Trail Blazers mounted an off-road expedition through the Everglades from Fort Myers to Miami. They would have starved if their Seminole guides hadn't provisioned them with freshly killed venison occasionally supplemented with airdrops of emergency rations. They finally emerged from the jungle, without their bogged-down vehicles, but with proof that you could complete a highway through the Everglades from Tampa to Miami. The nation had followed the Trail Blazers' progress through daily press releases and couldn't wait to follow them on the Tamiami Trail, to be completed only a few years later.[4]

Prior to the Trail Blazers' feat, Fort Myers had been a dusty old cattle town where cowboys herded the descendants of cattle released by Ponce de Leon onto special cattle boats bound for Cuba. Havana was their closest connection with civilization. For years, Henry Ford and Thomas Edison owned the only automobiles in the town. Henry Ford had had them shipped by boat to Fort Myers so the two friends could zoom up and down the well-worn oyster-shell trail between their beloved winter homes.[5]

On nearby Estero Island where conquistadors once careened their ships, the ex-rum runner Jack Dyle built a bridge so sports could drive their jalopies over the bridge from Fort Myers and onto the island's hard-packed sandy beaches. There they could enjoy all-night nude beach parties and bootleg booze at his beach-side casino.

But most of all, real estate was booming. Salesmen traveled through the Midwest on special trains extolling the Sunshine State as the nation's new center for year-round agriculture. The soil was so productive you didn't have to fertilize it. Never mind that most of the lots were still underwater and would have to be "reclaimed" from the Everglades. Many settlers had already built up agricultural communities on the Atlantic ridge running from the Loxahatchee River, north of Palm Beach, and onto Homestead, entranceway to the Florida Keys.

Speculators were so eager to get in on the boom that they literally threw checks at salesmen. In one auction they snapped up four hundred acres of mangrove swamp for $33 million in less than thirty minutes. Businessmen could stroll down Flagler Street buying a lot from a stranger on one block and selling it for a $10,000 profit on the next. A veteran discovered that the ten acres he had casually traded for his overcoat during the war were now worth over $10,000.[6]

Hollywood captured the era in two of its more madcap films. In *Some Like it Hot* a cross-dressing Jack Lemmon chases "Sugar," an often inebriated chanteuse played by Marilyn Monroe, as they hide out in one of Flagler's swanky hotels to avoid being rubbed out by Al Capone's hitmen. In *Cocoanuts*, Groucho Marx plays the role of a fast-talking land shark: "Can you get stucco? Boy can you ever get stucco!" In real life, bathtub gin sometimes sold for $38,000 a quart!

On Lake Okeechobee, Florida crackers were growing citrus, sugar cane, and vegetables behind an earthen levee built to reclaim the lake's soft black muck. But by the summer of 1926, heavy rains had raised the lake to the edge of that dike . . .

Meanwhile, back in Miami, Richard Gray issued a storm warning for the Great 1926 Miami Hurricane only hours before it would blow into town. And blow into town it did, packing 140 mile per hour gusts and the highest sustained winds ever yet recorded. A fifteen-foot storm surge swept through Coconut Grove, uprooting homes and floating them down the street. There hadn't been any severe hurricanes in recent memory, so when the storm's large eye passed over the city people rushed outside to celebrate. They had dodged another bullet. To his credit, Richard Gray threw open the doors of the weather station and yelled to people to get back inside.

"This is just a lull. The worst is yet to come."[7]

More than a hundred people ignored his pleas and were swept away by the eastern eye wall of the storm that engulfed them half an hour later.

After damaging every building in downtown Miami, the storm continued its march northwestward toward Lake Okeechobee. There, it blew a wall of water through the earthen dike into downtown Moore Haven and out across the flood plains. People were trapped in their beds as the lake surge burst through their windows and doors. One man grabbed his family and ran to higher ground, saving only his wife, three kids, and a soggy ten-dollar bill. In the end, the Great Miami Hurricane killed four hundred people and left forty thousand homeless. The man-made earthen levee had only concentrated nature's fury.

In 1927 the Mississippi overflowed its levees, killing 1,750 people and leaving 400,000 homeless.

In 1928, another 140 mile per hour hurricane flattened Puerto Rico as the island was celebrating San Felipe day. It churned on to destroy Palm Beach, then barreled into Okeechobee. This time the San Felipe–Okeechobee Hurricane blew a fifteen-foot lake surge through a twenty-one-mile hole in the newly rebuilt levee, drowning the farming communities of Miami Locks, South Bay, Chosen, Pahokee, and Belle Glade. In scenes eerily premonitory of 2005 New Orleans, people cut through their roofs to escape the flood and hung on to floating fence poles, tree trunks, and bloated cows. There they fought off the swarms of large, angry water moccasins also desperate to crawl onto dry land.

The San Felipe–Okeechobee Hurricane killed 2,500 people, mostly black sharecroppers working in their landlords' fields. Richard Gray had again predicted that South Florida would be spared, but it hadn't really mattered: None of the sharecroppers owned radios anyway. The San Felipe–Okeechobee Hurricane was the second most deadly hurricane in American history, drowning more people than Katrina in 2005.

Governor Martin counted 126 bodies along the six miles of road from Belle Grade to Pahokee. In another move that portended 2005 New Orleans, the Red Cross initially announced it would not rebuild the flood-prone homes where people had drowned, but it reversed the decision because of national pressure. Many of those same homes were flooded only a few years later.

If the Great Miami Hurricane occurred today, it would cause $90 billion in property damages, almost twice as much damage as caused by Hurricane Katrina. If the San Felipe–Okeechobee Hurricane occurred today it would cause millions more in damages and leave many more towns and neighborhoods drowned. Both areas have been developed far beyond their carrying capacity.[8]

Florida Today

On December 11, 2000, President Clinton signed an $8 billion bipartisan bill to revive the Everglades. It seemed to be an iconic moment, the signing of the largest environmental restoration project in history, a signal that we humans had reached a critical turning point in our relationship with the planet. We spent most of the twentieth century fighting nature, and now we would spend most of the twenty-first century repairing the damage.

As the congressmen and senators filed out of his office, Clinton turned to one of the legislative aides and joked, "If you don't do something about climate change, your Everglades is going to be underwater."[9]

How accurate was the President's offhand remark? The best way to find out is to approach Florida like a hurricane boring in from the Bahamas. Below you shimmer the shallow waters of the Bahama Banks, a vast expanse of shoals less than twenty feet deep. To the east is a ragged string of cays, the only remains of an island that used to be the size of Florida. Is this the future of Florida as well?

Florida's dominant feature is her extreme flatness. From Lake Okeechobee south to Florida Bay, a distance of over a hundred miles, the elevation drops only two inches per mile. The bottom third of Florida averages less than ten feet above sea level. The only moderately high land is the Atlantic Ridge, a sand-covered limestone spine that runs from Palm Beach to Miami. South of Miami the formation continues as the Florida Keys, a string of low islands that extend 140 miles to Dry Tortugas, but the ragged islands only stand about five feet above sea level.[10]

To put this into perspective, as I write this passage I am looking out over the Ipswich Marsh, forty miles north of Boston. Only one of our average New England high tides would leave a third of Florida underwater. That is also the same amount of marshlands that Katrina washed away in less than an hour.

The reason that Florida is so flat is because it spent most of its existence underwater. When the supercontinent Pangea rifted apart during the Triassic era, it left a raised plateau on the shoulder of the American plate. This became the underwater crucible in which Florida would form. Marine organisms thrived in this warm shallow sea. When the organisms died, their limey skeletons rained down on the ocean floor and became encrusted with more lime precipitating from the supersaturated waters above. These nucleated grains of calcium carbonate built up into thick beds of sediment called oolitic sand, because its grains look like billions of agglutinated fish eggs. Today, that oolitic sand has the scratchy feel you experience when you lie on some of Florida's beaches.[11]

However, the combined weight of these ever-growing beds of sediment was more than enough to cause the ancient sea floor to subside, even as more planktonic shells continued to rain down from overhead. This is why Florida is underlain with porous Biscayne limestone and has thick deposits of phosphate used to make fertilizer and gypsum used in drywall. Both are made up of the skeletal remains of these teeming marine creatures.

Florida only emerged from this watery crucible during the Ice Ages. It is estimated that the Florida platform flooded and emerged at least four times during the Ice Ages' last million years of glacial freezing and melting. Rising sea levels during the last interglacial warming period piled up great beds of oolitic sediments along Florida's leading eastern edge. Calcium carbonate in these marly limestone waters helped cement the sediments together to form the Atlantic Ridge, the raised spine that forms the only barrier between South Florida and the rising Atlantic.

When Paleo-Indians roamed early Florida only twelve thousand years ago, an eyeblink in geological time, Florida was almost twice as large as it is today. For the last five thousand years, however, average sea level has only risen about six inches every fifty years, not enough to cause catastrophe but enough to cause the kind of destruction I witnessed as a boy growing up on Cape Cod.

Now we have mounting evidence that the benign amount of sea-level rise we have enjoyed over the last five thousand years is about to increase exponentially. During the past decade, scientists in different fields have confirmed that carbon dioxide released by humans is warming the planet and that the warming has already caused less ice to form in the Arctic Ocean. However, two of the most alarming papers published in *Nature* in 2006 showed that some of Greenland's glaciers are retreating three times faster than ten years ago. If the entire Greenland ice sheet were to melt, it would raise the ocean's sea level twenty-three feet. If just a quarter of the Antarctic and Greenland glaciers melt it will raise sea level ten feet, and all of South Florida will be underwater. This is expected to happen in the next century. However, the flooding will not happen little by little over a hundred years. It will happen in short spurts during extreme storms like Hurricane Katrina, which destroyed the equivalent of all of the lowlands of South Florida in just a few short hours. When Florida does succumb to the rising Atlantic, all that will be left will be the remains of the East Coast's Atlantic Ridge, a string of ragged islands like the cays of today's Bahamas.

When I started working on this book I thought I would be writing about traditional coastal policies, things like deciding whether to use seawalls or beach nourishment to protect coastal homes. Gradually I realized that we are in a whole new ball game — all the rules have been changed. We are no longer discussing losing a few beach-front summer homes; we are looking at the destruction of major cities and areas like South Florida worth trillions of dollars.

This damage will probably happen whether we do anything about global warming or not. In the 1980s, scientists calculated that the entire planet would have to revert to using the amount of oil that only the United States used prior to World War II to affect the amount of warming we can expect from the amount of CO_2 already in the atmosphere and from the latent heat already trapped in ocean waters. And right now, humanity seems far more intent on using up as many resources as possible before somebody else gets their hands on them, than it is in conserving the few resources we have left.[12]

Florida shows us that although it is important to try to slow global warming, what we really have to start thinking about is how to adapt to the effects of global warming, things like more hurricanes and sea-level rise. Prior to Katrina, our first impulse after a major storm destroyed an area was to figure out how to rebuild it bigger and better than before. Now we have to start thinking of storms as helpful guides to show us where we should not rebuild and which projects should not be undertaken because they will be irreversibly destroyed in the future.

So was Clinton right? Does it make sense to spend an estimated $80 billion to revive the Everglades, or for that matter, $20 billion to restore coastal Louisiana? Both may be fully inundated in our children's lifetimes. Have we reached the tipping point where we should think about abandoning these areas? My heart tells me we should free the Mississippi so it can rebuild Louisiana's marshes and save the Everglades by allowing water to flow naturally as it did before. My head tells me it may already be too late to save these priceless natural resources—as well as major cities like Miami and New Orleans.

Cayo Costa, May 20, 2006

During the 1960s the National Park Service came up with the idea of creating a new category of parks called "national seashores." The primary purpose of national seashores was to protect major beaches for public use. However, their secondary role was to begin to create a national system of undeveloped barrier beaches to protect the mainland against storms and sea-level rise. This may become their most important function in the very near future.

The Gulf Islands National Seashore was supposed to be the crown jewel of this new system. Today it is the largest national seashore in the country, running 160 miles from Cat Island, Mississippi, to Santa Rosa, Florida. But if you look at a map, you will notice that there is a fifty-mile gap in the park. None of Alabama's fifty miles of barrier beaches are included in the seashore. Dauphin Island was the main reason for the gap. This also

THE GULF ISLANDS

R. C. Forget

shows how difficult it is to purchase barrier islands to protect the coast from sea-level rise.[13]

Dauphin Island became world famous after its picture was plastered on the covers of all the major newspapers in the wake of Hurricane Katrina. An aerial shot showed the twisted wreck of an offshore oil rig lying stranded on the beach; behind it lay the swept-over foundations of once elegant summer homes. Most of these buildings had been rebuilt and destroyed six times in the past ten years.[14] Their rebuilding was paid for by federal flood insurance subsidized by taxpayers—taxpayers like you and me.

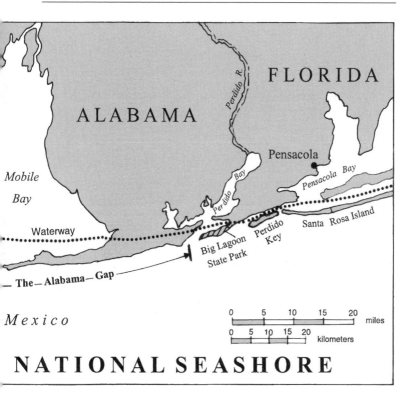

ALABAMA

FLORIDA

Pensacola

Mobile
Bay

Waterway

The — Alabama — Gap

Big Lagoon
State Park

Perdido
Key

Santa Rosa Island

Pensacola Bay

Mexico

| 0 | 5 | 10 | 15 | 20 | miles |
| 0 | 5 | 10 | 15 | 20 | kilometers |

NATIONAL SEASHORE

The island has looked this way for decades. The first time I visited Dauphin Island I was intrigued with its beach houses built on stilts over the Gulf of Mexico. I assumed the houses were designed that way so people could fish off their wrap-around porches. What I didn't realize was that the houses were originally built on dry land, then raised on stilts after one hurricane, then left abandoned in the Gulf of Mexico after subsequent storms washed the island out from under their foundations. Today the beach looks as natural as it has always been, but now it lies several hundred feet behind the houses.

The beach has retreated five hundred feet in the past twenty years!

All of Alabama's barrier beaches should have been prime candidates for inclusion in the Gulf Islands National Seashore. If they had been included, none of these homes would have been built and then destroyed by Hurricane Katrina and earlier storms.

However, politics has a way of intervening in these matters. During the 1930s, only 250 people lived on Dauphin Island, and most of its twenty-mile extent was owned by descendents of Frank Roykin and Forney Johnston. Today 1,500 people live on the island, but there are 3,000 absentee property owners.[15]

Since Hurricane Frederic in 1979, the federal and state governments have paid more than $90 million to rebuild the Dauphin Island Bridge and provided millions more dollars in federal flood insurance money to rebuild these mostly income-producing investment properties, after hurricanes Elena, Danny, Georges, and Ivan. Despite this governmental largesse, the island has largely failed as a public resort. Visitors who helped pay for the improvements through their taxes now cannot walk to the beach because they have to pass through gated communities and across private property—private homes that have been destroyed and rebuilt six times in the past decade at taxpayer expense.

But during the 1960s, most Alabama politicians considered the federal government to be the enemy, whether it was enforcing Civil Rights or creating national parks. They had the power and backing from wealthy homeowners to see that the forty miles of barrier beaches from Dauphin Island to Perdido Key were kept out of the national seashore. If these islands had been included in the Gulf Islands Park it would have prevented thousands of people from losing their homes and saved millions of dollars spent on rescue efforts, cleanup, and rebuilding.

Not far away, another island took a decidedly different course. Cayo Costa is one of a string of barrier beach islands that protects Charlotte Harbor, Pine Island Sound, and the Caloosahatchee River on Florida's fabled Gulf Coast. This area was the traditional center of the Calusa people, Native Americans who moved into this area sometime after the last Ice Age. The Calusas thrived in this warm, mild climate, becoming one of the only cultures in the world to develop large, permanent, year-round settlements not supported by food surpluses based on agriculture. Their secret lay in the warm shallow waters of the estuary, which provided them with an abundant source of high-protein seafood. In fact, our word for estuary comes from Estero Island, one of the southernmost islands in the barrier chain. The Calusas developed a highly organized society with complex social and religious structures. They continued to live safely on raised mounds in the mangrove islands behind the barrier beaches. Their religious and social leaders, called *caciques*, held sacrificial services on broad step pyramids surrounded with elaborate water courts, interconnected canals, and open chichee temples. The mounds were fortified by twelve-foot-high seawalls constructed with limestone marl and lightning whelks whose protruding white whorls made striking spiral patterns on the mosaic walls.[16]

When warriors spotted Spanish caravels approaching from the Gulf of Mexico, they were able to light signal fires and paddle through the system of canals and waterways to warn the villages of impending trouble. The islands became the Calusas' first line of defense against conquistadors and missionaries, working so well that the Calusas and their descendants became the only Native American group able to keep the European invaders out of their traditional territories. However, the mounds and seawalls provided another service as well. During hurricanes they provided protection against coastal flooding.

Unlike nearby Sanibel and Captiva Islands, no bridge was ever built to connect Cayo Costa to the mainland. This left the island able to fulfill its function as a moveable barrier against major storms, while people continued to live like the Calusas on the islands behind the barrier beaches. This made it easier for the state to purchase the almost uninhabited island and make it into a state park in 1971.[17]

The wisdom of that effort was tested when Hurricane Charley arrived on Friday, August 13, 2004. Charley took everyone by surprise. After passing over Key West as a Category 3 hurricane, it turned unexpectedly northeast and increased to a Category 4 storm in less than three hours. Some people speculate that when it made landfall directly on Cayo Costa it was actually a Category 5 hurricane packing 156 mile per hour winds. This was the first time since 1906 that two hurricanes, Bonnie and Charley, hit the same state within a twenty-four-hour period.[18]

Over a million people were evacuated, and Charley caused $15 billion in damages. It was also the second strongest hurricane since Hurricane Andrew in 1992. Three years later, storm evacuees still live in drug-infested FEMA trailer parks, in nearby Punta Gorda.

Nonetheless, Cayo Costa still looks much like it looked five hundred years ago. Tendrils of fog lift slowly off its nine miles of powdery white protected beaches. The invasive Australian pines, so unstable they toppled over in the first few minutes of Hurricane Charley, are mostly gone. State officials had spent years trying to remove the invasive trees with chain saws before Charley did the deed for them.

Now native sea grapes thrive in their newfound sun. Wild boar emerge from palmetto thickets to root for mole crabs and mollusks in the tidal flats. Feral horses nuzzle visiting boaters, who stay in simple cabins for only $30 a night. Alligators lurk in the shallow lagoon just behind the swimming beach, and the productive waters of the Gulf of Mexico produce billions of shells that wash up on the quiet beaches of Cayo Costa—beaches that still remain ready to absorb the power of the next hurricane to roar off the Gulf of Mexico. This is the benefit of leaving a barrier beach in its natural state.

"Our Town": Ipswich, Massachusetts, May 9–22, 2006
Halfway up the East Coast, my leisurely trip north was interrupted by a weather crisis at home. A stationary high sitting off the East Coast was blocking storms from marching across the continent and then drifting harmlessly out to sea. Instead, a giant line of storms was drawing water up from the Gulf of Mexico, then dumping it on New England. Every night on the Weather Channel I was mesmerized by arrows of rain sweeping in off the Atlantic directly onto Ipswich, Massachusetts. My hometown was about to become the epicenter for the floods of 2006.

Ipswich is a small town forty miles north of Boston. Our town only boasts nine thousand people, but our beaches and islands were the shooting location for both *The Crucible* and *The Thomas*

Crown Affair. To outsiders we are known mostly for our clams, our old homes, and our favorite son John Updike, who made a successful career for himself by chronicling the sexual peccadilloes of our leading citizens. He claimed he got all his best ideas sitting in church, and who am I to dispute him?

Driving down our narrow streets is a little like driving through a movie set. We have more small old colonial homes than any other community in the United States. One of our former houses was moved lock, stock, and barrel to Washington, D.C., to be an exhibit at the Smithsonian Institution. It is said that the reason we still have so many of our old homes is because we never had the money to remodel them, nor the good sense to tear them down.

Like most coastal New England towns, our central village is built around a river. During the eighteenth and nineteenth centuries the Ipswich River provided power to run small lace and textile mills. Today the old mills make trendy sites for restaurants, newsstands, museums, and art galleries. Artists at the art gallery openings will gleefully show you which page they appeared on in Updike's latest novel. A scattering of real estate and law offices fills in the rest of the sites. Our largest business is EBSCO, a high-tech publishing company that sits above the dam that used to power the eighteenth-century Ipswich Hosiery Company.

As I said, our town only boasts nine thousand people, mostly good people who volunteer on our town boards and support the local historical society. Many of them worked long hours to convince the town to build a new footbridge to tie together the two main streets that lie on either side of the river. Oh yes, I almost forgot: We have four old stone arch bridges in town, one of them the oldest in the country, and two newer nineteenth-century bridges up above the dam.

On May 9 everything changed. My flight from New York City was delayed by hurricane-force winds, and another plane was struck by lightning. Nobody seemed too concerned about the storm, just another northeaster that would eventually blow out to sea.

Things looked worse by Mother's Day. We had to use the old Mill Road to drive down to Boston, because all the main thoroughfares leading out of town were closed. Monday morning we woke to the unprecedented news that school would be closed on account of rain! The river was rising and most of the town's main streets were flooding.

School was canceled again on Tuesday, but the rain let up enough in the afternoon so we could all venture outside to see the damage. Two of the old stone bridges had to be closed, making the center of the town an island. You couldn't get there on foot or by car. A river, three feet deep, was rushing down South Main Street and into everyone's basements. The water was already neck high in the Italian restaurant. Its large wooden tables floated around the dining room. Ten million dollars worth of expensive books and pamphlets sloshed about in five feet of water in the EBSCO building.

The problem seemed to be that not-yet-opened footbridge. Its low metal frame was acting like a dam to divert water down South Main Street and into the EBSCO Building. Fortunately, it was May, so there was no ice to block the dam further. The police chief ordered officers to move gawkers off the Choate Bridge because the footbridge above it might give way and sweep them downstream. On the nearby Merrimac River, two pedestrians were seen clinging to a bridge that was being swept out to sea. Later we heard they had successfully swum to shore.

By May 22 it was mostly over. Eleven inches of rain had fallen in less than two weeks. The river had crested 10.67 feet above normal. One person died of a heart attack after his car stalled in rising waters in neighboring Topsfield, and another abandoned his car and was found drowned in a tangle of bushes only three hundred feet away. A third man made it to a shelter but was so cold from hypothermia he couldn't speak. The entire town heard the blast when lightning exploded a hundred-foot pine tree, scattering log-sized chunks of wood in all directions.

One of our bridges moved several inches, and people spotted water squirting out of the stone foundation of another. Our two upriver bridges would be out of commission indefinitely. Fifty percent of our businesses suffered major damage, and half of our homes had flooded basements. Thirty people cheered when a truck finally arrived with a special shipment of pumps sent directly from the factory in New Hampshire. EBSCO sustained $10 million in damages, including $300,000 a day in lost wages and revenues.[19]

But what I was really interested in was how our town would respond. Would we simply breath a sigh of relief and rebuild bigger and better than before, or would we learn from our mistakes?

The biggest mistake seemed to be the footbridge. Many people had worked extremely hard to obtain funding for the bridge, which would allow pedestrians to cross the river and sit in gazebos on attractive brick patios on either side of the river. But it seemed obvious that the footbridge was an engineering failure almost as significant to Ipswich as the levee failures had been to New Orleans. The reasons were basically the same. New Orleans' levees had been built to withstand a one-hundred-year storm based on a hundred years worth of data compiled in 1929! Our footbridge had been built to withstand a one-hundred-year storm based on a hundred years of data compiled in 1985.

However, everybody knows that hundred-year storms aren't what they used to be. I calculate that I have already lived through half a dozen "hundred-year storms" in my lifetime. Ipswich has had at least four one-hundred-year storms in the past six years. Last autumn, meteorologists concluded that based on world weather patterns we entered a new era of more intense storms starting in 1995. From now on insurance companies will base their premiums on this new data. If the town takes over ownership of the footbridge from the state, will it just be setting itself up for future liabilities? Will Ipswich become another irreversible mistake?

Ipswich also presented a golden opportunity to investigate how the federal flood insurance program really works. I started to make some calls. None of the affected businessmen had policies. More significantly, none of the lawyers, real estate agents, or insurance company owners could tell me how the program worked or why they didn't have policies. Here was a program that offered artificially low premiums specially designed to protect people in flood-prone areas, but even the best informed businesspeople couldn't tell me why they didn't own the insurance.

I soon discovered that this has always been the problem with the federal flood insurance program. Premiums from many policyholders are supposed to cover those few who get in trouble in any particular year. At the same time, the program is supposed to protect taxpayers from having to bail out communities after every flood. But the program has never worked because most people don't realize they are eligible for the inexpensive insurance.

The only time people hear about flood insurance is when they apply for a mortgage, and they discover they have to buy the mandatory federal flood insurance because their house is in a one-hundred-year floodplain. When this happens, most people simply turn around and hire a local engineer to take elevations to dispute the flood maps. However, it usually costs far more to fight the added cost than to simply buy the inexpensive insurance. Homeowners generally pay $300 to $400 a year for a federal flood insurance policy that provides them with $250,000 in coverage. So our Italian restaurant would have only paid $1,800 over six years for coverage and would potentially been reimbursed half a million dollars for its two "hundred-year" loses in 2001 and 2006.

But the main reason most people don't take advantage of federal flood insurance is that in their heart of hearts they don't really believe they are going to need it. Floods are only something that happens to people in other places like Florida and New Orleans, not on a peaceful river in New England! The result is that the federal flood insurance program only takes in $2.2 billion per year in premiums. Last year it had to "borrow" money from taxpayers to pay for the $25 billion in claims from Katrina, and the program was already $2 billion in the hole from former storms. The borrowed money will have to be paid back with interest.

This still didn't explain why nobody in Ipswich had the inexpensive policies. Little did we know that while we were cleaning up after the storm, a Senate subcommittee was marking up a bill to try to rectify that situation. The process had started in the fall of 2005 when Barney Frank, one of our congressmen from Massachusetts, collaborated in introducing a bill to increase the number of federal flood insurance premium holders by expanding flood zones from areas affected by a one-hundred-year storm to areas affected by a five-hundred-year storm. Increasing the size of flood zone areas would bring millions more people into the mandatory program, both insuring them and making the program more fiscally viable.

What happened? Nothing! Lobbyists jumped all over the bill. Real estate agents said the bill would increase the cost of buying homes, and homebuilders argued that it would ruin the economy because people might stop building in flood zones. But wouldn't most people think not building in a flood zone is a pretty good idea? Actually, most people never heard about the bill. Lobbyists went to Congressman Gary Miller, a former homebuilder from California, who quietly made an amendment to table the bill for further study.

It turned out that FEMA actually didn't have any decent maps of floods caused by five-hundred-year storms. So Congress took several steps back, and while we were drying out our basements on Sunday, May 21, committee members marked up a new bill to remove the exemption for mandatory flood insurance for houses and businesses built behind dams and levees. It had taken several days of telephone calls to legislative aides and Senate subcommittee members, but I finally had the answer to why no one in Ipswich had federal flood insurance. We lived below a dam—a dam that had "protected" us almost as well as the levees had "protected" New Orleans!

If downtown Ipswich had been protected by the federal flood insurance program, what would have happened? Our Italian restaurant probably would have been raised eleven feet above the one-hundred-year flood level to conform to the federal flood insurance program's mandatory building codes. Its customers would have only been allowed to dine upstairs or on open outside patios. The restaurant would have saved itself tens of thousands of dollars in rebuilding costs. EBSCO would have been required to move its publications to a warehouse on higher ground. Businesspeople would have been required to suspend their furnaces from basement ceilings and move their kitchens and electric panels to higher floors. The town and state would have avoided embarrassment by building its footbridge with a graceful arch high enough to avoid a five-hundred-year storm. Taxpayers would not have been asked to put up millions more dollars in disaster relief to help our town. Not a bad outcome for a simple piece of legislation that had been stalled in Congress because of a group of wrongheaded but highly effective lobbyists.

Legalities: Topsail Island, North Carolina, June 5, 2006
On May 22, 2006, the National Oceanic and Atmospheric Administration, ominously referred to as "NOAH," announced its annual forecast for the 2006 hurricane season. It predicted thirteen to sixteen named tropical storms would form in the At-

lantic, and of those, eight to ten were likely to become hurricanes, including six Katrina-sized major storms. Nobody gave much credence to those numbers. For the preceding three years NOAA had underforecast hurricanes. In 2005 it predicted only eight hurricanes, and fifteen had formed, including the three deadly sisters Katrina, Rita, and Wilma.[20]

On May 23, Al Gore released *An Inconvenient Truth*, an unlikely blockbuster based on his slide show about global warming. For the preceding thirty years while we, his environmental colleagues, were safely ensconced in our ivory towers, he battled global warming in the political arena. Roundly ridiculed as the "ozone man" by the former President Bush, he lost the White House to the junior President Bush and the Supreme Court in 2000. However, Al Gore's film signaled that global warming was finally going mainstream. It was about time. Many of us had seen the same slides as undergraduates and had naively assumed the issue would be quickly addressed. Instead, faith-based politicians and conservative talk-show hosts had outmaneuvered legitimate climate scientists, labeling them as a bunch of whacko Cassandras trying to take away America's God-given right to drive gas guzzlers and consume 75 percent of the world's resources. Oil companies like Exxon-Mobil supported as many as forty think tanks whose sole purpose was to spread misinformation under the guise of scientific research. Journalists fell into their sophist trap by giving these petroleum-supported spokespeople equal billing with serious scientists.

On June 1, the first day of the 2006 hurricane season, the Army Corps of Engineers released its nine-volume IPET report that revealed the Corps' many failings in New Orleans. The main problem was that the Corps had planned for a "standard hurricane" derived from models. However the "standard hurricane" had turned out to neither as powerful nor complex as the real thing. The Corps' engineers did not take into account that land under New Orleans' levees would subside, so that when Katrina struck the levees were three feet lower than they should have been. Plus, the Corps' flood control system contained no redundancy, so when one part started to fail it led to a cascade of further damage.

Most significantly, the report chronicled a series of plans that were overruled and replaced with fallback systems that ultimately proved to be the city's undoing. After the Corps' proposal for a storm surge barrier was turned down for environmental reasons in the 1970s, the Corps had proposed building floodgates on the city's drainage canals instead. When the floodgates proposal was blocked by local levee boards, the Corps retreated again to propose building floodwalls along the canals. It was these floodwalls that failed so catastrophically during Katrina.

On June 2, we people in Ipswich received our own unwelcome surprise. Just when we thought we had dodged a bullet, Massachusetts reclosed our three main bridges. This meant the town would be restricted to one bridge for the indefinite future. Businesses that had been badly damaged were going to be cut off from their customers during the crucial upcoming tourist season. Perhaps we should have expected something like this. During the storm, lightning had struck the same street twice, hitting a house only blocks from the pine tree that had been struck the week before.

The entire episode had been strangely unsettling. There had been no hurricane, no wind damage, no major storm to interrupt daily life, just intermittent rain that wouldn't go away. The rain raised a primal fear. What if the world had changed from a nurturing garden of Eden to a place of malevolent evil? Was this the beginning of a new seasonal pattern? The previous spring we had experienced a similar hundred-year northeaster that had flattened our barrier beach, and weather forecasters were troubled by a warm-water ring that had spun off the Gulf Stream and was now lurking 120 miles south of Nantucket. Would its hot waters kick start this summer's New England hurricanes the way the Gulf of Mexico's hot waters had kick started Katrina? Were we entering another unanticipated weather pattern brought on by global warming? Although our planet has experienced similar climatic changes throughout its history, this is the first time human civilization has ever had to deal with such drastic change. Human civilization was born in a crucible of climatic stability. Our little rainstorm showed just how disruptive such seemingly minor changes can be.

But people on Topsail Island in North Carolina were thinking of other things as the hurricane season approached. Like most coastal areas, the Carolinas' Outer Banks have a colorful history. Pirates and blockade runners used these barrier islands as havens during war and as wrecking stations during intermittent times of peace.

You can still get something for nothing on the Outer Banks, but you have to be lucky. One of the luckiest men on Topsail Island had been an unlikely speculator called Mr. Boryk. M. A. Boryk was a fourth-grade-educated son of Ukrainian immigrants who used to love to row his skiff over to Topsail Island to fish. After the army abandoned part of its old Topsail Island missile testing site in the 1960s, Mr. Boryk and two friends rowed over to the island to make a bid on twenty-five acres of the newly available land. They won and proceeded to draw subdivision lots out of Mr. Boryk's leather cap. M. A. ended up with the southern tip of the twenty-six-mile-long island not far from Camp Lejeune.

During the intervening years, Mr. Boryk grew rich developing his share of the 25 acres. But that was not the full extent of Mr. Boryk's luck. During those same intervening years the rising Atlantic eroded seven miles off the northern end of Topsail Island and delivered it via longshore currents to the southern end. This increased the original 25-acre plot by 125 acres—125 acres that included a mile of oceanfront property worth more than $30 million![21]

Hundreds of thousands of other people bought land on barrier islands during the same period of low hurricane activity that ran from the 1960s to the 1980s. Flood insurance was available, so why not build? If your house got knocked down, you could always collect and build again. The federal flood insurance program has paid more than $500 million to repair homes in North Carolina since hurricanes increased in 1995. In the old days, only poor people lived on places like Monomoy and Topsail Island. Today, thanks to insurance, only rich people live on the coast.

Most of the sinking northern end of Topsail Island is only habitable because state and local government rebuilt the infrastructure after Hurricane Fran in 1996 and Hurricane Floyd in 1997. But such largess may be coming to an end. When Mr. Boryck's daughter, Annette Appegaard, informed the town she wanted to develop her new 125 acres, Topsail Island's mayor saw red. Mayor Parrish only oversees an annual town budget of $1.7 million dollars. He couldn't afford to continue paying to repair roads and utilities after every storm. The Topsail Planning Board agreed and strengthened its land use designation for the tip of Topsail Island from primarily conservation to strictly conservation. Their new designation would effectively block Mrs. Appegaard from developing her land, but it also raised a constitutional question. In 1972 the Supreme Court decided that when the South Carolina Coastal Commission changed the law so that David Lucas could not develop his land, it constituted a "taking" under the Fifth Amendment of the Constitution. So Mrs. Appegaard, like Mr. Lucas, should be reimbursed for the value of her land. Topsail backed down. It could not afford the $30 million.

Instead, the Topsail Island township asked the North Carolina Land Trust to apply for a grant to purchase the land. They would not know the fate of their grant until well after the end of the 2006 hurricane season. (As of this writing local state and federal agencies are still trying to raise all the necessary funds.) But Mrs. Appegaard was willing to wait. "The Lord giveth and the Lord taketh," she told a visiting reporter for the *Wall Street Journal*.[22]

Not a bad trick, getting paid $30 million for land that you never really owned and that had been created by the effects of sea-level rise. But $30 million will still be far less than the potentially billions of dollars in damages that could be done to the infrastructure and a hundred new twelve-room beach-house McMansions in future storms. These are the types of decisions that towns, states, and federal agencies will have to continue to make as we filter the effects of more hurricanes and sea-level rise through the laws, institutions, and economies of our form of liberal democracy.

The Ash Wednesday Storm, New Jersey, March 5–8, 1962
Anyone interested in the history of our coasts should spend some time looking down on New Jersey, preferably from a very high altitude. I had this opportunity on a flight up the East Coast. It was on one of those crystal clear, cobalt blue days when the beaches stand out in stark contrast to the dark green waters of the Atlantic. I could see all of New Jersey's oceanfront cities as well as the latticework of rusting groins, bulkheads, and seawalls built to prevent those cities from washing away with the ocean.

Cape May stuck out below us like a sore thumb throbbing above Delaware Bay. Cape May is our oldest coastal resort, established by New England whalers in 1692.[23] However, things didn't pick up appreciably until railroads started bringing tourists to Cape May's famous beaches in 1850. Soon presidents, robber barons, and movie stars were frolicking on her wide beaches, and John Phillip Sousa was giving concerts on summer evenings. But today Cape May's famous beach is gone, thanks to a massive seawall built to prevent erosion, and the city has lapsed into a quiet senescence.

Next, the streets of Atlantic City stood out below me. There were Broadway, and Park Place, Ventnor, and St. Charles. The inventor of Monopoly expropriated the names of Atlantic City for his popular board game that celebrates the irrational exuberance of a former real estate boom. I could just see the Reading and Pennsylvania railways. In 1856 they had transformed this desolate barrier beach into the hottest real estate market in the world. I could almost see Donald Trump entering his opulent Taj Mahal casino. Today Atlantic City's famous beach is also gone and the state has pledged to pay $5 billion over fifty years to renourish its sand: not a bad way for "the Donald" to protect his assets, but at a pretty high cost to you and me.

Finally there is Ocean City, with the highest annual coastal engineering budget in the nation. So if you want to defeat a foolish bit of coastal development, remind people of the New Jersification of this coast and most will understand. The Garden State has almost as many jetties, groins, and seawalls as coastal homes. A recent beach renourishment project buried unexploded ordinance in the new sand. New Jersey has made every coastal mistake.[24]

But New Jersey met its match during the 1962 Ash Wednesday Storm. It was not a hurricane, but the storm proved to be a catalyst for our present-day understanding of barrier beach dynamics. The rogue storm arrived on a moonless night in March, then raged up and down a thousand miles of coast for three more days. It struck when the moon was closest to the earth and the sun and the moon were aligned. These astronomical conditions produced the feared perigean spring tides that only occur every two years. The storm also struck after winter storms had already removed much of the beaches' protective sands, and it continued through five cycles of maximum high tide erosion. It was the worst possible storm, occurring at the worst possible time.

The storm devastated the coast from Florida to Cape Cod, but slammed into New Jersey particularly hard. Thirty-foot waves crumpled Atlantic City's famous steel pier and splintered Ocean

City's boardwalk. Forty-five thousand beachfront homes tumbled into the Atlantic in New Jersey alone.

However, the day after the storm passed, coastal dwellers were stunned by the damage. Scores of inlets had slashed through barrier beaches and the dunes had been all but flattened. Such storms are not known for killing people. But this one had killed thirty-two people and caused half a billion dollars in damages. Perhaps It stood out particularly clearly in people's minds precisely because it occurred in an era of fewer hurricanes when people were less accustomed to experiencing major storms.

If the Ash Wednesday Storm occurred today, the sea level would be five inches higher, the waves would be five times more powerful, and the ocean would inundate the land a hundred times further inland. A thousand miles of the East Coast would look like the beaches of the Indian Ocean after the 2004 tsunami.

But coastal geologists noticed something particularly interesting. The undeveloped islands were able to slough off the worst effects of the storm and after a few weeks they had already started to recover. That summer, waves pushed the displaced sand back onto the islands from the offshore bars and filled in most of the new inlets. By autumn, beach grass had revegetated the most eroded areas of the beach and the dunes had started to reform. Officials were so impressed with the resiliency of the natural system in Virginia that they canceled plans to build 45,000 private beach homes and created the Assateague National Seashore instead.[25]

The Ash Wednesday Storm arrived at a crucial moment in America's evolving conservation ethic. In September 1962, Rachel Carson ushered in the modern environmental movement with the publication of her blockbuster, *Silent Spring.* The book not only indicted the pesticide industry, it also questioned the bedrock of America's long-standing belief in the traditional concept of progress, the idea that technology and money should be used to fight and defeat nature. Yet the battered beaches of New Jersey showed that something was wrong with that traditional ethic. All the expensive groins, jetties, and seawalls had not prevented the damage. In fact, like pesticides, they seemed to have made the situation worse. Yet nobody knew exactly why that was so. Here was a stable beach. You should have been able to armor it with seawalls to prevent erosion. What was wrong?

Three years later we started to find some answers to the question when James Keeling published a paper that showed that carbon dioxide had risen precipitously ever year since atmospheric scientists had started measuring it in 1957. It was clear from the data that increased carbon dioxide would lead to global warming, and global warming would lead to more coastal erosion and faster sea-level rise.[26]

Keeling's data also challenged coastal geologists' old concept of erosion. Traditionally they had assumed that barrier islands formed several thousand years ago in approximately the same place you find them today, so it made sense to try to stop them from eroding.

Operating under that old paradigm, the Civilian Conservation Corps built a hundred-mile-long sand dune along North Carolina's Outer Banks islands during the Depression. But by the 1960s the National Park Service noticed that the islands were eroding on both their front and back sides. It made sense that the ocean would erode the fronts of the islands, but why were the backs of the islands receding as well?

It is rumored among coastal geologists that it was an inebriated old Outer Banks hermit who first came up with the answer. He kept showing up at public meetings insisting that the islands were not eroding, just migrating. Almost everyone laughed at his silly rantings, but the notion still niggled at the back of some scientific minds.

Eventually the National Park Service sent a young graduate student into the field to investigate. Paul Godfrey discovered that storms normally wash sand off the front of barrier islands and deposit it on their back-side marshes in a process called rollover. But the artificial sand dune was so high that it was preventing sand from washing over the island, so the back-side marshes and beaches were receding from lack of sand. The town crank was right, barrier islands do migrate by rolling over, and anything you do to prevent that process is doomed to failure. But the problem is that the islands only migrate episodically during storms, so it is easy to ignore the long-term pattern of inevitable migration, especially if you are a developer intent on building a beachfront home.[27]

Coastal geologists now realize that most of the East Coast and Gulf Coast barrier islands formed on the edge of the continental shelves and have migrated to their present positions as the sea levels have risen since the last Ice Age. In fact, the beaches, dunes, sandbars, and mainland all migrate across the continental shelf as part of an integrated system. The Outer Banks have migrated as much as fifty miles and many Gulf Coast islands have migrated more than a hundred miles toward the shore.

Some islands off Mississippi have migrated more than a hundred miles in the past century alone.

This new realization pitted coastal geologists against home-owners, who wanted to see each storm as an individual episode rather than an ongoing process, and coastal engineers, who wanted to continue getting paid. But there was still a problem. Nobody knew exactly how much and how fast the seas were going to rise because of global warming.

This was where the climate scientists made a big mistake. In 1980, Stephen Schneider came out with a paper that predicted that global warming would cause the sea level to rise twenty-eight feet in the next hundred years. Journalists scooped up the report, reprinting it with flashy graphics that showed the Statue of Liberty up to her keister in the Atlantic. But the media paid little attention when Schneider quietly retracted his numbers in an obscure footnote on bottom of one of the back pages of *Scientific American* nine years later.[28]

But the damage had been done. A generation of coastal geologists had grown up quoting the alarmist numbers. They had cried "wolf," and their cries would come back to haunt them. When coastal geologists codified their new thinking in the second Skidaway Statement, it landed on President Ronald Reagan's desk with a resounding thud. He was in the process of ushering in his own genial era of head-in-the sand anti-environmentalism.[29]

Developers were able to dismiss the Skidaway School of coastal geologists as a bunch of Cassandras who loved public beaches but had no concern for the lives of the people they wished to remove from the coasts. As the most articulate and visible environmentalist, Al Gore was on the receiving end of the worst of the later frat-house vitriol. His mentor Roger Revelle had been a prime mover behind Keeling's original research on climate change.

So, like the research on global warming itself, most of the science behind sea-level rise was done thirty years ago, then promptly ignored by policy makers. Today we know that the sea is rising at least six inches every fifty years and that this translates into a one-hundred- to one-thousand-foot horizontal retreat. This rate of sea-level rise will undoubtedly increase dramatically in the years ahead. Yet communities continue to develop the coasts as if sea-level rise does not exist, and as if barrier islands are stable entities instead of moving features on our rapidly changing planet.

The Big Apple: The 1938 Hurricane, New York
June 2006 continued with more disturbing weather. This time the torrential rains flooded the Northeast from Virginia to New York. Four children playing near swollen streams in Virginia and Maryland were swept to their deaths. Two drivers near Binghamton, New York, drowned when two bridges and an entire section of Interstate 88 washed away. In the end, nine people died and ten New York counties were declared disaster areas. The only bright spot was that the IRS building in Washington, D.C., was closed for several days—almost as good as a no-school day!

The rain was caused by the same conditions that had caused our Mother's Day floods in New England. The Gulf Stream was still spinning warm core eddies off its northern edge and was starting to spew a five-hundred-mile swath of eighty-degree water into the western Atlantic. The abnormally hot waters now splayed offshore from North Carolina to well beyond Bermuda. It was almost as if cooling waters sinking off Iceland had stopped pulling the Gulf Stream so far north. Instead the Gulf Stream waters were just spilling out haphazardly into the Atlantic. If so, this would indicate the beginning of global warning's most frightening scenario, the breakdown of the North Atlantic Conveyor that drives the Gulf Stream. Meteorologists speculated that the northerly postion of the Gulf Stream was responsible for blocking the El Niño fronts, causing our never-ending spats of rainfall. But what was more disturbing was that this swath of eighty-degree water could work like the warm waters of the Gulf of Mexico to kick start hurricanes and accelerate more powerful storms directly into New York City.

After New Orleans and Miami, New York is generally considered to be the third most vulnerable city to heavy damage from a major hurricane. But insurance analysts point out that after Florida, New York has more insured coastal property than any other state. In the case of a direct hit, New York might be most vulnerable, Florida second, and New Orleans a paltry fifth.

Meteorologists know that New York is due for a major hurricane. It has not had one since 1938. They calculate that there is a 99 percent probability that New York will be hit by a tropical storm in the next fifty years and a 73 percent probability that it will be hit by a major hurricane. What will happen? The 1938 hurricane can give us some answers.[30]

On September 16, 1938, a Brazilian freighter reported a severe storm off Florida. It had developed largely unnoticed off the coast of Africa and had been nourished by the unusually hot waters of the Atlantic. By the time the U.S. Weather Bureau heard about the storm, it was only able to issue a hurried warning for Miami. During the night, however, the storm unexpectedly swerved north, missing the Southeast mainland altogether. It was curving around a "Bermuda high," so the weather service assumed it would continue to blow safely out to sea. However, the Bermuda high had itself moved north, funneling the hurricane over the Gulf Stream. There, it sucked up more energy and gained rapid forward speed, becoming "The Long Island Express," the fastest moving hurricane in recorded history.[31]

It had been 117 years since the last hurricane hit New York. That was in 1821, when the Hudson and East Rivers had merged across Manhattan as far north as Canal Street. But nobody thought it could happen again, so no warnings were issued for the impending disaster.

The sky continued to grow dark by the afternoon of September 21. Telephone poles were snapping like matchsticks on Long Island, but hundreds of people continued to stand on the beach, transfixed by a low-lying bank of fog rolling slowly toward them. A few wise souls decided to retreat across the only remaining open bridge, but most did not. They were swept away, as the fog bank proved to be not fog at all, but the rapidly moving gray waters of a thirty-foot storm surge.[32]

Although the storm passed fifty-five miles east of New York City, wind readings at the top of the Empire State Building were 120 miles per hour, twice as high as at street level. Storm surges killed hundreds more people and swept homes off their foundations without a trace in Watch Hill, Rhode Island. The surge then pushed jumbles of cars and boats into downtown Providence, leaving the city under thirty feet of sea water and piles of debris.

From Rhode Island, the storm continued north, destroying the magnificent old fisheries building in Woods Hole, and sweeping up the Connecticut River Valley and even toppling a trestle of the famous old cog railway on the summit of New Hampshire's Mount Washington. Eventually the storm swept through Montreal and finally petered out over the cool tundra of the Arctic. Behind it stretched a nine-hundred-mile trail of destruction through six states and the Canadian province of Quebec. The hurricane had killed six hundred people and caused $300 million in damages.[33]

There was relief in New York City. Flooding had shorted out electricity in most of Manhattan and in all of the Bronx. But it had been a near miss. The Big Apple had dodged a bullet.

What would happen if the 1938 hurricane hit today? It would be the sixth costliest hurricane in history, causing $18 billion in damages. But what would happen if the Big Apple sustained a direct hit? At least $300 billion in damages, almost twice as much as New Orleans!

New York City has several unique features that make it particularly vulnerable to a major hurricane. First is its size. Officials would have to find ways to help more than two million people evacuate the city. Both the George Washington and Verrazanno Narrows bridges are so high that, like the Empire State Building, they would experience stronger winds than at street level, requiring them to be closed several hours before the storm actually arrived. The same is true for the two ferry services that operate across Long Island Sound. They would also have to be shut down six to twelve hours before the storm surge entered the sound. Evacuation would have to be initiated as much as day earlier than for other cities, which would be complicated by the fact that northern hurricanes move faster than southern ones.[34]

Both LaGuardia and JFK airports would flood under twenty feet of sea water. JFK was built on the remains of Hog Island, a gilded seaside resort that had been swept away by the hurricane of 1821.

Sea water would pour down subways and into the Holland and Brooklyn Battery tunnels, stranding passengers aboard trains as happened when the weakened remains of Hurricane Dennis blew through in 2004. In all, over seventy square miles of the area, including the entire financial district, would be under water. The effects of the closure would reverberate through the world's economy for months, if not years.

The Army Corps of Engineers does not estimate lost lives, but in a 1990 report on New York it stated that "Storms that would present little to moderate hazards in other regions would cause heavy loss of life in New York City."[35]

It is estimated that over a million people would be without power for two weeks to a month after the storm. As in New Orleans, it would take people years to recover from their losses.

Is there anything that can be done to prevent this tragedy? Malcolm Bowman, a physical oceanographer, believes there is. He and his colleagues at the State University of New York (SUNY) have proposed building four-mile-long storm-surge barriers that could be rotated up to create a wall three feet above the normal high tide level. These would protect Jersey City, Manhattan, Jamaica Bay, and JFK airport but would leave parts of Brooklyn and most of Long Island fully exposed. The project would cost tens of billions of dollars and be paid for by both those to be protected and those to be left exposed, not a particularly strong selling point for a project that would require broad political support.[36]

Dr. Bowman has already applied to the state for a $3 million grant to conduct a feasibility study. But the project is not likely to happen anytime soon. If there is one thing we have learned about flood protection, it is that it usually takes a catastrophe to initiate any meaningful action.

Such a catastrophe occurred half a century ago in England, when a 1953 storm pushed high tides up the Thames, drowning three hundred people, destroying farmland, and generally scaring the pants off of Savile Row Londoners. Construction was finally started on the Thames Barrier in 1982, thirty years after the tragedy.

The same 1953 storm affected the Netherlands as well. It breached the famous Dutch system of dikes in 450 places. Sea water and ice flooded into cities and towns, drowning 1,900 people as they slept, and leaving cows to die a slow death from drowning or hypothermia. The captain of a ship in Rotterdam saw the problem and immediately sank his ship in a breach in the dam to protect the city. The Netherlands engineering response was equally heroic.[37]

Holland constructed a $3 billion system of forty-foot high dikes, man-made islands, and flood gates that control both the entry and exit of sea water. However, the country had no choice. It is called "the Lowlands" for a very good reason: More than half of the Netherlands lies twenty-four feet below sea level. Plus, the flood-prone Rhine, Maas, and Schelde Rivers also flow through its leveed countryside. Holland's natural coast was shortened four hundred miles by the elaborate system, which was built to withstand a 10,000-year storm, instead of a hundred-year storm as required in the United States.

Unfortunately, 10,000-year storms are also no longer what they used to be. If Greenland and Antarctica continue to melt as quickly as scientists now expect, the $3 billion Dutch system may be sorely tested within the next fifty years. A similar tidal barrier system in Venice is already obsolete because of sea-level rise.

It is doubtful that New York will decide to build such a barrier system without a major storm as catalyst. So, if such engineering solutions are too expensive, what does New York plan to do? Maryann Marrocolo, the city's assistant commissioner for emergency planning, explains, "Our basic philosophy is that there's very little we are going to able to do to stop the surge. What we want to do to minimize the damage is to raise public awareness and plan for evacuation and recovery efforts."[38]

A Tale of Two Cities: Galveston and Indianola, the 1900 Storm

Before ending this grand tour of the coasts, there is one last place and one last storm to visit. Though the storm occurred on the Texas coast more than one hundred years ago, it has particular relevance for New Orleans today, and for the rest of our hurricane prone coasts tomorrow.

Galveston sits on a thirty-mile-long sandbar not much wider and almost as vulnerable as Monomoy, the island I used to visit as a boy on Cape Cod. The main difference is that a deep channel sweeps around the easternmost point of the island and on into a natural deep-water harbor on Galveston Bay. Despite the

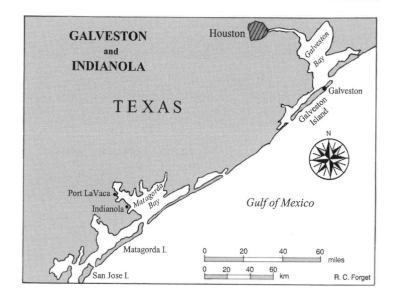

GALVESTON
and
INDIANOLA

TEXAS

Houston

Galveston Bay

Galveston

Galveston Island

N

Port LaVaca

Indianola

Matagorda Bay

Gulf of Mexico

Matagorda I.

San Jose I.

0   20   40   60
miles

0   20   40   60
km

R. C. Forget

vulnerability of the island, the location of the harbor was just too good not to be turned into a major port.

Barges and later railroads carried cotton, wheat, and cattle down the bay and across the causeway to the city's bustling wharfs. There, the well-muscled stevedores used two-hundred-pound jackscrews to cram bales of loosely packed cotton into the holds of oceangoing ships. A good team of screwmen could increase a cargo by as much as 20 percent. They were so valuable to wealthy white shipowners that Norris Wright Cuney was able to organize an all-black Cotton Jammers Association that wielded considerable political power. Cuney became a force in the city and was elected as the city's first black alderman. White businessmen sought his counsel in negotiating deals fair enough to blacks to keep the port humming and harmonious.[39]

As in many old cities, Galveston's landowner and business classes had largely pulled away from city government. One exception was the Deep Water Committee, established in 1881 to bring James Eads to Galveston to make a case for building jetties to scour the channel, as he had done so successfully in Louisiana. After the Galveston channel was deepened, however, the group of influential bankers and businessmen continued to lobby for better city governance. Many of their ideas formed the core of the sweeping municipal changes that would occur in the wake of the upcoming storm.

Prior to the storm, women lacked the vote and seemed content to stay at home, organizing the many picnics, dances, concerts, and bicycle races that made Galveston such a fashionable seaside resort. Indeed, almost everyone in the city did their best to ignore the fact that they lived on a hurricane-prone, two- to three-mile-wide barrier beach island, which at its highest elevation was only nine feet above sea level. That should have inspired about as much confidence as living twenty feet below sea level in New Orleans. In fact, seven major hurricanes had ravaged this East Texas coast from 1835 to 1900. Two of them had successfully wiped out Indianola, Galveston's main seaport rival, a hundred miles to the south.

In 1874, Indianola was the second largest port in Texas after Galveston. One of its most exotic imports had been two shiploads of camels that the Secretary of War had transported to Indianola to help move military equipment through the Southwest. Other than that, Indianola exported cattle and was the favored entry port for European immigrants traveling to West Texas.

But Indianola's fortunes changed drastically after a powerful hurricane washed over the low-lying city in 1875. Although most of the city's structures were eventually rebuilt, the underpinnings of the economy were swept away and the port slipped into economic stagnation. When a second, more powerful hurricane struck eleven years later, dispirited investors rerouted the railroads, and the town moved ten miles north to Lavaca. Today, most of the former city lies under water, a victim of erosion and rising seas. All that remains is a granite marker placed on the shore at the point nearest to Indianola's old courthouse, now three hundred feet offshore in Matagorda Bay.[40]

A few forward-thinking people in Galveston used the demise of Indianola to resurrect the old idea of building a seawall to protect their port city. But after a few hurricane seasons passed with no appreciable storms the idea quietly died, as it had so many times before. Instead, people seemed content to continue lopping off the top of the dune that ran like a spine down the center of the island. They used the sand to fill in the island's low-lying areas to squeeze in more homes, and they cut the island's erosion-resistant cedar trees to give people easier access to the beach.

When the cotton season opened on September 1, 1900, Galveston was the largest city in Texas and boasted the highest per-capita income in the state. It had just eclipsed New Orleans as the nation's largest exporter of cotton and was behind only New York City in exports of wheat.[41]

September 8, the day of the Galveston Hurricane, dawned warm and overcast. The live oaks in Kempner Park swayed in the tropical breezes blowing in quietly off the nearby Gulf of Mexico. Visitors lounged on the broad verandas of the Tremont

Hotel or strolled along the city's beaches, so wide they hosted automobile races in season. But one of the people on the beach was there for a different reason. He was Isaac Cline, head of the Galveston Weather Bureau. Cline had the idea that high tides could presage a hurricane, and this tide was one of the highest on record. After taking his measurements, he had seen enough. He hitched up his buggy and drove down the beach, warning visitors to take the causeway back to the mainland and warning inhabitants to move to safe houses on Broadway, the city's main street that perched on the duneline. But Galvestonians mostly ignored his warnings. They had seen such hurricanes before; water would overflow into the gulf-side streets, then drain back after the storm had passed. Why should they leave the safety of their own homes? But by late afternoon the bay waters and the gulf waters merged in the center of town. Galveston was already submerged when the full fury of the storm descended on its inhabitants.[42]

By early evening a thousand people sought shelter in the Tremont Hotel. They rushed up the stairs to the hotel's mezzanine floor as water seeped into the lobby. Shortly after 6 P.M. the storm surge swept over the island. The water, which was already fifteen feet deep in Cline's house, rose four feet in an instant. Thirty-foot waves on top of the storm surge knocked gulf-side houses off their foundations and used them as battering rams to smash the next line of buildings. Survivors heard the cries of people as they were swept down the streets and out into the Gulf of Mexico. Many would be consumed by sharks in the days ahead.

By the end of the night, between six thousand and ten thousand people had drowned or been crushed by waterborne debris. The following day, Captain Thornton sailed a rescue boat down Galveston Bay. There were so many bodies in the water he had to send a man forward with a pike to push them away from the bow. All that could be heard in the strange calm after the storm was the quiet sounds of bodies bumping against the side of the boat as it drifted in the outgoing tide.[43]

It had been the worst natural disaster in North American history. Over a fifth of the city's inhabitants were dead. Whole neighborhoods were swept away, leaving only sodden sections of marshy soil behind. In the end, everything came to rest in a berm of rubble thirty feet high and three miles long. More than four thousand buildings, two thirds of the city's structures, were destroyed. Whites dragooned blacks into piling bodies onto barges and shipping them offshore to be dumped. Even though the corpses were weighed down with iron fastenings, many drifted the eighteen miles back to shore. Their remains had to be collected again and burned on funeral pyres that lit up the beach for several months.

But it is what happened after the storm that should concern us today. The nation was in the midst of the Progressive Era. There was optimism in the air that science and engineering could solve almost any structural problem and that governmental reform could improve people's health and economic well-being.

The day after the hurricane, Mayor Walter C. Jones created the Central Relief Committee for Galveston Sufferers. The CRC, as it came to be known, included many of the progressive-leaning members of the old Deep Water Committee. The CRC established subcommittees to bury the dead, handle public safety, distribute relief, and initiate a bond drive to pay for the city's rebuilding. It did such an exemplary job that the city ended up in better financial shape than before the storm. The CRC also became the model for a new progressive system of city government run by seven commissions rather than by a mayor and twelve aldermen. Such ideas had been bandied about before, but it had taken the shock of the storm to turn them into reality.

On September 17, another force of nature blew into town. It was Clara Barton, the seventy-eight-year-old founder of the American Red Cross. She arrived in Galveston after traveling by train from Washington, D.C. It had been an arduous trip. Her train had snaked its way south to Atlanta, then west to New Orleans and on through to Galveston. She was so ill when she arrived that she retired straight to a bed in the Tremont Hotel.

The next day she started directing the relief efforts from her bedside table, ignoring pleas from Red Cross officials that she return to Washington for the sake of her health. She continued in this way for many months, raising funds, writing her own thank-you notes, attending meetings, and organizing ward relief committees. At her insistence, each ward committee was chaired by both a man and woman. Eventually the entire relief operation was turned over to Barton's direction, and she was appointed as the sole female member of the Central Relief Committee.

Perhaps Barton's longest lasting contribution was to draw so many local women into Galveston's political process. At the peak of her relief efforts she had 150 women and 50 men working on her ward committees. The experience they gained helped open doors to women in other civic organizations. One was the Women's Health Protective Association, which worked to improve the city's sanitation and notoriously bacteria-filled milk supply. The association also lobbied the state to jettison Galveston's old form of city government in favor of the progressive idea of a city run by commissions. The down side of all of these activities was that while middle-class white women gained access, they tended to displace black men in positions of political power. Norris Wright Cuney was sorely missed; he had died shortly before the storm. There is concern that the same dynamic could occur in New Orleans.

On September 25, 1901, the newly installed Commission Government appointed a committee to hire engineers to come up with a plan for protecting the city. The engineers had to decide which areas could be saved and which were irretrievably lost. Instead of fighting nature along the island's entire thirty miles of waterfront, they decided to build a three-mile-long, seventeen-foot-high seawall to protect the central city. Of greater importance, they initiated an audacious plan to raise the entire city seventeen feet.

First, every house, hospital, hotel, and cathedral was put on beams and raised seventeen feet in the air by human-powered jackscrews. Then dredges, steaming in and out of canals specially dug through the center of the city, pumped a slurry of sand off the floor of the Gulf of Mexico and under each building.

Throughout the twelve years of construction, people navigated the city by walking on an intricate system of rickety raised catwalks. By 1912 the project was complete, and Galveston was once again the second largest port in the country and the largest city in Texas. However, this status was not destined to last.

The federal government had expanded so much by this time

that it could finance the dredging of channels into cities that lacked natural harbors. One of these cities was Houston, Galveston's dreaded rival to the north. In 1914 the Army Corps of Engineers widened and deepened Houston's Buffalo Bayou so oceangoing ships could enter its port. Galveston's fate was sealed. Houston had more room to build large wharves and port facilities. It was ready to accommodate the offshore petroleum industry when it took off with the discovery of the Spindletop oil field. Houston roared past Galveston to become the center of the nation's oil industry, the largest and most influential city in Texas, and the fourth most populous city in the United States.[44]

Galveston's seawall finally got its test when a second, equally powerful hurricane slammed into the city in 1915. Almost all of the structures not protected by the seawall were destroyed, and the streets were flooded, but the city survived.

After the storm passed, however, people realized that high-energy waves slamming against the seawall had washed the beach entirely away. Since then, the city has had to pump sand back onto the beach on a regular basis and construct an expensive system of groins to slow the sand from washing away. This is the reason that seawalls have largely gone out of favor as a way of preserving coastal land, plus the fact that in 2007 it would cost $30 million per mile to build such a seawall!

Today, however, Galveston is once again a thriving seaport and a favorite stop for cruise ships. Shops and mansions dating from the city's prehurricane Victorian era still line its crowded Broadway shopping district. The city hosts museums, hospitals, and the country's first nonmilitary level 4 biocontainment facility. In case of another hurricane the facility would be instantly decommissioned and all the anthrax, Ebola, and West Nile viruses destroyed.[45] *Fail Safe* meets *The Perfect Storm*, doesn't that make you feel more secure?

But in the end you would have to say that the seawall worked for Galveston in 1900. Though it no longer had the power and ambitions it harbored before the storm, Galveston emerged with a smaller footprint but a safer infrastructure and a better, though less representative, system of government. The essence of the city had survived. Will the same thing happen in New Orleans?

Beach Renourishment: Instant Gratification?

If seawalls are too expensive and cause almost as much erosion as they prevent, are there other ways to protect a coast? Another barrier beach island on Florida's Gulf Coast provides an answer. Gasparilla Island is half a mile wide and seven miles long. Celebrities as disparate as Harrison Ford and the entire Bush clan stay at the Gasparilla Inn, and Jimmie Buffet often visits his sister at the south end of the island.

A string of tasteful old Gasparilla homes line the Gulf of Mexico shore. Most of the houses languish behind a ten-foot seawall. For decades if you wanted to swim in the Gulf you had to clamber down the seawall and swim between rusty old groins. Most owners preferred to swim in their own private pools. Anglers cleaned their catch on the seawall and threw scraps to the sharks that regularly patrolled the runnels between the offshore bars. Each wave would dislodge teeming colonies of mole crabs and coquinas that dug furiously to rebury themselves before the next wash carried them offshore. Every spring, sea turtles laid their eggs in the powdery white sand.

All that changed in 2006. After seven years of lobbying, Lee County officials finally persuaded Congress to fund a $13 million beach renourishment project. Renourishment has become engineers' favored "soft solution" to deal with coastal erosion. During the 1970s and 1980s coastal geologists demonstarted that "hard solutions" like seawalls created more erosion than they prevented because they scoured away the beach in front of seawalls. But there are problems with "soft solutions" as well.

I was able to visit Gasparilla Island halfway through a beach renourishment project in 2007. The experience allowed me to tally up some of the pluses and minuses of the procedure. The biggest plus was the beach itself. You no longer needed to clamber down the ten-foot seawall. You merely stepped over the top of the seawall and onto the brand new beach. But a significant short-term minus was the rusty, four-foot-diameter pipeline that emerged from the Gulf, then snaked its way two miles down the beach. All night you could see the lights of the offshore barge sucking up tons of sand, shells, and water from Boca Grande Pass and pumping them through the submerged pipe to the shore. All day long you could hear the quiet scraping and scratching of the shells as they sluiced through the exposed pipe in a slurry of murky gray water.

A mile down the beach, workers diverted the pipeline into two pipes lying parallel on the beach. A valve switched the slurry into the offshore pipe and a thirty-foot gray geyser of water, sand, and shells shot down the beach. After the new sand formed a peninsula running parallel to the shore, the engineers switched the flow to the inland pipe, which blasted more sand to fill in the area behind the peninsula. After the entire segment of beach was buried eight feet deep in sand, bulldozers flattened the beach and new pipes were added to the ever-growing pipeline. In this manner, the project moved down the coast, adding three hundred feet of new beach every week.

A year ago, sharks and dolphins swam in what was now a three-mile-long, eight-foot-deep berm of gray shelly sand. It would take five years for the mole crabs, coquinas, and other invertebrates to restablish themselves. But it didn't look like the beach would last even that long. The night after we arrived, a cold front swept in off the Gulf of Mexico and six-foot waves battered the new beach. In a matter of hours the storm removed twenty feet off the face of the beach. The equivalent of several football fields filled eight feet deep with sand had been washed away.

Cold-front storms occur frequently along this coast. In less than six months most of this sand will have washed away, and only twenty feet of beach will remain. This looked like a modern version of King Canute's attempt to thwart the tides. But looks can be deceiving. After the storm passed, my wife and I were able to snorkle offshore. Gentle waves had created a foot-high ledge of shells along the edge of the beach. We were able to swim along this underwater ledge collecting the best shells as they tumbled down the face of the miniature cliff. But once we swam beyond the small ledge, we were swimming over a shallow plateau of new soft sand. The waves had already winnowed most of the shells out of the sand. The theory behind this project is that most of the sand will stay in a three-mile-long offshore cell of sand, and when major storms attack, the underwater sand will absorb the waves' energy to prevent damage to seaside homes and condominiums.

However, there are always problems when theory runs into reality. The project had originally called for the construction of an offshore breakwater and two T-groins to contain the sand within a three-mile-long sand cell. Unfortunately, bids for the structures had come in at double what they had been budgeted for, so the project proceeded without them. This allowed storms to push sand back into the depths of Boca Grande Pass, losing it to the offshore sand cell. Plus, an auxillary pump broke down two months into the project, threatening to extend the project's completion until well into the turtle nesting season.

But the most ludicrous situation arose as homeowners realized that members of the public might actually want to use the beach their taxes had paid for. This meant they would also want to park on the shell-lined roads that led to the beach. But the homeowners had come to think of both the roads and the beach as their own private property. When I arrived, residents were frantically planting twenty-foot-high palm trees on both sides of the roads to prevent parking. Evidently the homeowners would rather lose their homes than have the public exercise its right to use the roads and beach.

With all the glitches, it is doubtful that the Gasparilla project will remain effective for its projected seven years. Even if it does, Congress will have to continue spending $13 million every seven years to renourish this coast. At that rate, it will cost about $4 million every seven years to protect a mile of this coast. Should the public pay so much money to defend private beach-front homes? That is the ultimate problem with beach renourishment: It costs a lot of money for what is ultimately a short-term, Band-Aid solution to coastal erosion.

The other problem is sea-level rise. While we were snorkeling my wife discovered a large lightning whelk. The side of the whelk's shell contained a curious three-quarter-inch hole. It was only after we returned home that we realized that the hole had been chipped into the side of the shell to hold the end of a wooden handle. My wife had discovered an ancient Calusa axe.

Several hundred years ago, when Gasparilla Island had been a tenth of a mile further out to sea, the area we had been swimming over had been a Calusa village. This island, like all barrier beach islands, has been retreating since the end of the last Ice Age. A $13 million beach renourishment program is not going to stop the effects of sea-level rise.

Chapter 4

The Future

Rearranging Deck Chairs?: July 2006

On July 17, 2006, a massive plate of the earth's crust suddenly lurched toward the earth's interior, creating an offshore earthquake and tsunami that sent forty thousand people scurrying for high ground and four hundred dead bodies rolling in the incoming tides of Indonesia. It was a reprise of the 2004 tsunami that killed 170,000 people in nearby Banda Eceh. The cause had been the same, a new cycle of increased geological activity that has made earthquakes, volcanoes, and tsunamis more frequent. Indonesians would have to decide again whether to run for the hills or rebuild in the face of the new reality. It is the same decision Americans will have to make in the face of more frequent storms.[1]

Change was making itself felt on other parts of the globe as well. The first half of 2006 was the hottest six months in recorded history. Eighty people died from heatstroke in Europe, reviving the specter of the 2003 heat wave that killed thirty thousand elderly people. Cyclists in the Tour de France were sweltering through the 116-mile Alpine stage, while London broiled in hundred-degree weather.[2]

In the United States, Midwest farmers sifted powdery dry soil through their fingers and sprayed their cattle with water to try to keep them alive. Two large amorphous blobs of triple-digit air straddled the country from southern California to the Midwest. The rest of the country simmered in ninety-degree weather. Storms and hot weather left 200,000 without electricity in St. Louis, 100,000 in Queens, and 17,000 in Southern California, where 132 people died and the temperature reached a record-setting 126 degrees in Death Valley and 116 in Palm Springs. Twenty-five thousand cattle died because they lack sweat glands to rid their huge bodies of heat. Walnuts were cooking in their shells in the Central Valley. And this was just the beginning of the hottest summer in four hundred years.[3]

The conditions were also ripe for hurricanes, but so far El Niño–generated winds had been strong enough over hurricane breeding sites to shear the tops off of any nascent storms before they had a chance to become fully organized. However, this had not stopped coastal newspapers from fanning hurricane hysteria. Anxiety was rippling up and down the nation's coasts. Florida homeowners watched their insurance premiums double to $13,000 per year for a 1,500-square-foot building, and on Cape Cod it was no longer possible to purchase commercial hurricane insurance for private homes. For the first time in more than a hundred years, more people were moving out of Florida than moving in. Part of the reason was that older people couldn't afford the higher insurance rates and were tired of dealing with the threat of impending hurricanes.

Along the Gulf Coast, oil companies still tried to repair the $31 billion in damages they had sustained in hurricanes Rita,

Katrina, and even Ivan from 2004. Fifteen percent of the companies' production was still off-line and drivers had watched gas prices rise toward $4 per gallon in the United States and $8 per gallon in Europe. Chevron quietly sank its Typhoon platform into the Gulf of Mexico, deciding it was cheaper to sink the multi-million-dollar structure than to repair it. Five hundred workers were working week-long shifts to repair the giant Mars platform, then being flown back by helicopter to spend an off week on cruise ships doubling as temporary floating hotels.

New Orleans had become a city of disturbing contrasts. When the AFL-CIO announced it would invest a billion dollars to build affordable homes and a new hotel in New Orleans, its move was greeted with good cheer. Billions of dollars of federal aid were also starting to stream into the city. More houses were sold for more money than before Hurricane Katrina. Most of the action was occurring along the "Sliver on the River," the old part of town that had not been flooded when the levees broke. But bargain hunters were also scooping up flooded houses for half of what they had sold for before the storm. Real estate agents earned more in the first six months of 2006 than in all of 2005.[4]

Things were not so rosy in formerly flooded black neighborhoods. Homeless people still camped along narrow strips of brush and trees growing on the sides of the Mississippi. More lived in cars abandoned underneath major highways. Electricity had still not been restored to many parts of the city, and even where it was available it cost twice as much as before Katrina. Power outages were common, and people retreated to their front stoops to sweat through the heat without air conditioning or even so much as a glass of cool water. Sleep was impossible. Residents trudged through their days feeling tired, bitter, and constantly depressed. Was it really worth the effort to try to raise children under such conditions?

People still pulled over to the side of the road to sob at the sight of what had happened to their beautiful city. These were the classic symptoms of posttraumatic syndrome, but this was trauma that wouldn't go away. You had to face it again and again, every day.

Many people had done the emotional calculus and simply decided they didn't want to continue living in such a pervasive atmosphere of constant depression. They were the silent statistics, people quietly dribbling unnoticed out of the city. Others took more drastic action, like a young black man who simply walked into the churning waters of the Mississippi hoping to drown. The suicide rate was close to triple what it had been before the storm.[5]

Louisiana and Mississippi were in the midst of a huge experiment in social engineering. The U.S. government was in the process of distributing $10 billion directly into the hands of individual homeowners. It was an unprecedented amount of relief money. One of the goals of the experiment was to see whether people would decide to head to higher ground or relocate in areas almost certain to be devastated by future storms.

However, each state had a slightly different take on the experiment. Louisiana wanted its citizens to rebuild, so it devised a system of incentives so that homeowners would only receive the full value of their home if they rebuilt, but 40 percent less if they moved out of state. That had been the traditional approach taken by other states struck by similar disasters.

The state of Mississippi took the opposite approach. It wanted to discourage people from rebuilding houses on the coast, where they would be likely to be destroyed by future storms. But instead of passing restrictive environmental regulations that could run into eminent domain problems and be overturned in court, the state decided to let people weigh the consequences and decide for themselves. People would be allowed to use the full amount of their compensation money to either rebuild or relocate, but homeowners could also use the full amount of their compensation for an entirely different purpose, like sending their kids to college or replenishing their savings accounts depleted by the storm. The state would benefit from not having to pay for future rescue operations and not having to rebuild infrastructure in flood-prone areas. Most homeowners realized this was their last chance. They would not be offered such a generous deal if they rebuilt and their homes were destroyed in a future storm.[6]

Which state is taking the better approach? It depends on your assumptions. Is Katrina a one-time event that will not be repeated? Is New Orleans on the verge a dramatic turnaround, or will it continue the gradual decline it was in before Katrina? Will basic services like electricity and water be restored at reasonable prices? Will the levees hold?

In all fairness, the two states are in different situations. Things are far more straightforward in Mississippi. Most of the destroyed houses there were destroyed by storm surge. It was clear that the homes never should have been built there in the first place and that they are susceptible to future storms. Most of the houses destroyed in New Orleans, however, were only indirectly damaged by storm surge, but were directly damaged by humanly caused errors in levee construction, errors that can potentially be fixed.

Congress has decided that New Orleans is vital to the economic well-being of the country, has more historic buildings than any other equivalent American city, and plays a unique role in our cultural heritage. It is as valuable and unique as London or Venice and therefore should be saved and rebuilt. Which state's approach is more progressive in the face of the present reality? Only time and future hurricanes will tell.

Monday Night Football: The Case for Optimism,
September 24, 2006

September 24, 2006, was another iconic night for New Orleans. It was the night the New Orleans Saints returned to the Superdome for the first time since Hurricane Katrina. One of the commentators made the point that FEMA money had helped pay for the rebuilding, which seemed fitting. New Orleans was not going to recover without income, and what better income to have than from food-loving, Bourbon Street–bound football fans?

There were also reminders that some of the fans had been in the Superdome when it had been a "shelter of last resort" and that many were still living in trailers. But the dominant message was that the levees had been rebuilt, another Katrina had not hit, and the Saints were marching in—the Big Easy was back in business, NOLA was back in town.[7]

The message certainly resonated as a background story for a feel-good night of football, but if New Orleans is truly going to be saved, rebuilding must be looked at in an entirely different way. Instead of starting with the Superdome or the Lower Ninth

Ward, we must start by looking at the coast on which New Orleans lies.

At first blush it does not look good. New Orleans is surrounded on three sides by water at sea level, while the city itself lies in a flood-prone bowl seventeen feet below the Gulf of Mexico. Plus, New Orleans sits twenty feet below the Mississippi River, which is itself both three feet above sea level and a hundred feet below sea level where it flows through the city. The marshes, which could have saved New Orleans from Katrina's storm surge if they had been allowed to grow naturally for the past seventy-five years, are fraying at the rate of twenty-four square miles per year or almost a football field every three minutes. You can go out for a day of fishing and return to find that part of your yard has slipped underwater or your dock has tumbled into the bayou. Over the past twenty-five years, 20 percent of all the nation's repeat losses paid for by national flood insurance have occurred in Louisiana's Orleans and Jefferson parishes.

But there is also hope. If we first stand back and look at the coast, we see that there is great redundancy already built into the natural systems that protect New Orleans. There are barrier islands, inlets, living marshes, and natural ridges. All of these can be enhanced by undoing almost three hundred years of mistakes. The river is the key to city's protection. The most significant long-term improvement would be to divert the potential 400 million tons of sediment that used to flow down the river back onto the marshes and barrier islands. This would mean not just standing back and looking at the coast but also standing back to look at making agricultural and navigational changes in the entire Mississippi watershed. These would include removing dams in upstream states that prevent much of the Mississippi's potential sediment load from moving downstream and reducing the use of fertilizers so they don't make the marshes eutrophic and continue to pollute the Delaware-sized area of the Gulf of Mexico known as the Death Zone.[8]

Another hopeful feature of the marsh is that it is a living, growing entity. If the oil companies stop channelizing the marshes and the Third Delta Conveyance Project proceeds, the marsh grasses will be able to incorporate these sediments into an interlocking matrix of roots and peat. This living soil will be able to expand enough in five years to protect New Orleans against the amount of storm surge encountered in Katrina and enough in twenty years to dampen down the storm surge from a Category 5 hurricane. We will have created an eighty-mile expanse of living, growing, low-maintenance soil, a natural horizontal levee to protect New Orleans.

Storm surges could also be blocked by building barriers at strategic openings like the Rigolet's Pass into Lake Pontchartrain. This would be like a general deploying soldiers at a mountain pass, rather than stretching them along a long, vulnerable front, which is what engineers did when they built levees between the southern coast of Lake Pontchartrain and the city. Cypress trees could also be planted in Lake Pontchartrain to provide natural buffers to the artificial levees topped with canebrake plantings of American bamboo. The shallow roots of this native species would bind up the soil to help prevent the kind of levee failure that occurred during Katrina. Only after all these natural defenses have been considered and mapped should planners start looking at where development can be adequately protected.

Far away from the glare of the Superdome lights, there was evidence that the country was finally going to make some of these changes. The Bush administration had only been able to hold down the pressure for change for so long. Global warming, rising gasoline prices, and Hurricane Katrina released the power of our still resourceful and innovative nation. It was almost as if you could hear the gears of a new paradigm as they meshed into place. You could feel voters demand action and towns, states, businesses, and individuals take the initiative into their own hands.

Farmers planted corn for ethanol, towns built wind turbines, companies developed new technologies, and voters demanded that the federal government require higher mileage automobiles. The National Academy of Sciences released a report that suggested that states replace towns in making long-term decisions about coastal development and urged people to grow marshes instead of building seawalls to protect their shorefront homes. Even the insurance industry started to offer incentives for people to buy hybrid cars and to use green technology to rebuild their damaged homes.[9]

Nowhere was this new paradigm felt more than in the normally well-insulated walls of the Army Corps of Engineers. Katrina had spawned a year of internal debate. The old guard knew it had been caught with its pants down, and the new guard knew that if it was going to survive it had to try new approaches. They incorporated their new way of thinking into the Corps' IPET report released in June 1. In the exhaustive report, the Corps indicated it had to turn to other places and people to implement these new approaches. One place was the Netherlands, which had its own extensive history with floods and engineering. One person was a young American soil scientist who had started her own company because none of the existing engineering firms were willing to consider using natural processes in battling flood control.

Now one of the largest civil engineering projects ever awarded was about to fall on the shoulders of this woman, who had made it her life's mission to overturn the old school of engineering based solely on massively built hard structures like levees and seawalls, turning to one based on working with nature's natural redundancy to create multiple layers of protection.

Wendi Goldsmith arrived at her approach through a long process of self-directed education. At Yale she had studied geology and geophysics and spent so much time hanging around Yale's renowned School of Forestry that she described herself as their adopted pet mascot. From Yale she had gone on to study soil science, botany, and landscape engineering in graduate school before signing on as an apprentice to her mentor Lothar Bestman in North Germany. From here she had been able to return to her roots in nearby Holland. Both her maternal grandfather and great-grandfather had left Rotterdam to join the Dutch engineering team that helped build New York's Holland Tunnel in the 1930s. Later, on an exchange program at the Delft Hydraulic Institute, she witnessed the turnaround in the Netherlands' own thinking about flood control.[10]

There had been a lot to learn. As early as 1300, the Dutch had built their first community windmill to pump water from behind the modest dikes used to protect their homes and churches, which were themselves built on mounds to protect against floods. Today Rotterdam, the second largest seaport in the world, lies twenty-three feet below sea level and Amsterdam, the capital of Holland, lies twelve feet below sea level. In fact, 70 percent of the country's gross domestic product is produced below sea level. As the Dutch like to say, while God created the earth, the Dutch created the Netherlands.[11]

Like Louisiana, the Netherlands has also had a long history of natural disasters. The Dutch suffered seventeen major floods between 1700 and 1950. But in 1953, Holland experienced what it called its perfect storm. Hurricane-force winds and record high storm surges burst through the dikes in mid-December, killing 1,800 people, inundating 50,000 homes, and leaving 350,000 acres totally flooded. The United States supplied pumps and helicopters to help the Dutch dewater the country, so when Katrina hit the Dutch were quick to show their gratitude. They dispatched the royal frigate *Van Amsel* to rescue survivors in Biloxi and flew pumps, helicopters, and engineers to New Orleans to help dewater that city.[12]

After the 1953 storm, the Dutch vowed that no such catastrophe would ever happen again. They embarked on a national program to build dikes massive enough to withstand a ten-thousand-year storm, a level of protection that far exceeds that for a Category 5 hurricane. This was hard engineering on an unprecedented scale. But almost from the beginning, problems arose. Once-thriving estuaries became fetid, algae-clogged, freshwater holding ponds, and navigation suffered when sedimentation clogged former shipping lanes.

Slowly, the country started to modify its approach. It built moveable barriers that allowed tidal waters to flow freely during normal times but could be raised during storms to block storm surges. It changed zoning regulations so government had the power to prevent people from building in flood-prone areas. It set aside those areas to provide room so the country's three major rivers, the Rhine, the Meusse, and the Scheldt, could flow into their natural floodplains during storms without destroying homes and businesses. The Dutch built secondary channels to bypass their vulnerable urban areas. It has been a long evolution, but since the 1970s, the Netherlands' major preoccupation has been to learn how to manage both water behavior and human behavior, rather than to simply fight floods with massive engineering.[13]

After completing her European education, Goldsmith returned to the United States to interview for jobs, but none of the engineering firms were interested in doing the kind of bioengineering she wanted to pursue. So instead of working for someone else, she had started her own company, the Bioengineering Group, in Salem, Massachusetts.

As soon as Katrina hit, Goldsmith realized that New Orleans presented the ideal situation in which to use bioengineering principles to provide the best long-term protection for the city. She immediately started to shuttle back and forth to New Orleans, Washington, and Boulder, Colorado, to explain the virtues and procedures of bioengineering. In Boulder she reconnected with Arcadis, the American subsidiary of the Dutch firm that had designed the storm-surge barriers in Amsterdam, Rotterdam, London, Venice, and St. Petersburg. The company had "written the book" on coastal engineering and authored the renaissance in the new Dutch way of thinking about flood control.

Goldsmith's talks were less like a sales pitch than a delightful, wide-ranging college colloquium. She used her knowledge of geology and geophysics to provide the broad picture and her training in soil science, botany, and landscape engineering to provide details. She knew her audience well enough, and was herself enough of a realist, to know you had to use some traditional hard solutions to provide short-term solutions, but she never lost sight of bioengineering's long-term benefits. She almost tapped out all her personal and company resources to do it, but over several months she gradually won over more and more converts, who introduced her to others. Eventually, crusty old engineers from the traditional school of engineering within the Corps came around to her way of thinking, and cynics from the new guard, who had formerly despaired of ever seeing real change, realized she might just have the optimism and determination to see this through.[14]

One by one, traditional old engineering firms like Halliburton fell out of the running, and Goldsmith's joint partnership with the Dutch subsidiary Arcadis in Boulder and her startup bioengineering group in Massachusetts won the largest civil engineering contract ever awarded. But the real significance of the contract was that it showed that the nation and the Army Corps of Engineers had made a sincere commitment to a new greener way of thinking about coastal policy in our modern era of more powerful storms and sea-level rise.

Just Seconds from the Ocean: Codfish Park, Nantucket
Near the anniversary of Hurricane Katrina, I found myself in the seaside village of Sconset on Nantucket Island. We were staying

in a delightful summer cottage only fifty feet from the beach. But it gradually dawned on me that my affinity for this place represented the crux of the problem. People are drawn to the edge — the closer to the mists and the sound of the crashing ocean, the better. But while this cottage is the first place you would want to rent, it is the last place you would want to own.

The cottage is in Codfish Park, a neat little community of fifty cottages nestled below the Sconset Bluffs. On a beautiful summer's day it seemed like the perfect place to buy. In fact, the converted barn was for sale for almost a million dollars. Out of curiosity, I called a broker, who extolled the virtues of the unique community. What she failed to mention was that only fifteen years ago several of these same cottages has been swept out to sea by what locals called the "no name storm." The Weather Channel had filmed one of the cottages momentarily drifting in the tempestuous seas before a single wave exploded it into a thousand pieces of drifting wood.

But today there were no empty lots, no evidence of loss, just a cluster of charming cottages only steps from the beach. A broker might be excused for not bringing up the past. The first time the new homeowner would probably hear about the problem was when a neighbor knocked on the door requesting his $500,000 contribution to a community effort to pump offshore sand onto the beach, an operation that would have to be repeated every five to seven years. Without such beach renourishment, these cottages are but a single storm from oblivion.

The history of this area is revealing. Any surfer who has spent time looking for the new surf break can tell you that the beach changes every year. During the 1830s, ocean waves pounded directly on the bluff behind our cottage; our lot was underwater. Then, as so often happens on high-energy beaches, the locus of erosion changed, and a broad expanse of sandbar grew below the bank. The first people to take advantage of the new situation were Nantucket fishermen who built simple shacks on the beach where they stayed all summer, while their families enjoyed the more cosmopolitan pleasures of the town of Nantucket, still wealthy from the whaling trade.

Then, around the turn of the century, a celebrated group of Broadway stars developed an artists' colony on the bluff. They hosted such luminaries as Rosalind Russell, Bette Davis, Truman Capote, and John Steinbeck. Robert Benchley was the stalwart on the bluff. He had started the Round Table at New York's Algonquin Club, but was known best on the island for his recipe for soaking bluefish in gin to burn off its oils. His son wrote a film about a Soviet submarine that throws an island into a tizzy after running aground on a Sconset bar; his grandson wrote a film about a shark that throws an island into a tizzy after lunching on several of its summer inhabitants. When you think about it, *The Russians Are Coming! The Russians Are Coming!* and *Jaws* are the same stories with different antagonists.[15]

The fun-loving "up-bank" artists spent all summer amusing themselves with tennis, musicals, picnics, and the Sconset casino. Their servants, mostly black, settled into the former fishing shacks in erosion-prone Codfish Park. Today a few of the original "down-bank" cottages are still owned by the original black families, the rest by wealthy summer folk.

Today Sconset's main concern continues to be erosion. In 1991 the same "no name storm" washed seventeen feet off the top of

Sankaty Head. Multi-million-dollar homes hang over the edge of the bluff, only a single storm or four calm years away from tumbling down the face of the hundred-foot cliff. I took a photo of one of the houses hanging over the cliff, and a passerby suggested the caption should read, "Just seconds from the ocean." One of the "up-bank" Sankaty Head owners has spent more than a million dollars and eight years trying to fortify his sandy bluff. His neighbors have mostly opted to move their houses across the street and to the back of their once extensive lots. Meanwhile, a citizens' group hopes to move the historic Sankaty Lighthouse back to a neighboring golf course, but it only has a year before advancing erosion will make this impossible.[16]

It is clear that this small community has gone through some irreversible changes. Its citizens still try to recover from a storm that occurred fifteen years ago and erosion that threatens today. If I were a betting man, I would put my money on the people moving their homes rather than on their neighbor spending a million bucks to battle the rising Atlantic Ocean.

Meanwhile, New Orleans finally received a much-needed dose of reality from its newly appointed black recovery chief Edward J. Blakeley. In March 2007, the outspoken planner with impeccable credentials dramatically changed the rules. Homeowners would no longer be required to use their federal grants to rebuild their damaged homes—they could spend the grants any way they chose. This meant they could buy or repair safer homes in Houston or Atlanta, where they had already lain down roots, rather than being forced to rebuild their homes in New Orleans.[17]

The new rule will allow New Orleans officials to concentrate on rebuilding a smaller, safer city, but the ruling also indicates that the United States finally understands that we have irrevocably entered the new age of global warming, with rising seas and more intense hurricanes. More coastal cities like New Orleans and more areas like the Gulf Coast will be destroyed. If we don't have the vision and tools to deal with this new era, nature in the form of huge impersonal storms will step in to do it for us. In the face of this crisis, can we afford to continue building cities below sea level, and houses "just seconds from the ocean"? Much as I would like to stay in a place like Codfish Park, I know it is time to go.

Notes

Preface (pages ix–xiv)

1. William Sargent, "New Orleans: Rebuild Some Neighborhoods Inland." *Cape Cod Times* 9 October 2005.
2. Colby Strong, "Gulf Coast Researchers Assessment of the Psychological Impact of Hurricane Katrina." *Neuropsychiatry Reviews* 7.4, April 2006.
3. Cornelia Dean, "Expert Federal Panel Urges New Look at Land Use Along Coasts in Effort to Reduce Erosion." *New York Times* 13 October 2006.
4. Quoted in Philip Staples, "No More Excuses on Climate Change." *Financial Times* 31 October 2006.
5. Ibid.

Introduction (pages 1–6)

1. "Monomoy Island Wildlife Refuge." U.S. Fish and Wildlife Service web site, http://monomoy.fws.gov/html.
2. Beth Daley, "The Shifting Sands of Cape Cod's New Landscape Threatens Refuge." *Boston Globe* 17 November 2006, A1.

Chapter 1 (pages 7–20)

1. *Storms of 1954.* The Green Line, Canadian Hurricane Centre. http://www.atl.ec.gc.ca/weather/hurricane/html.
2. Ibid.
3. Kenneth Chang, "In Study, A History Lesson on the Costs of Hurricanes." *New York Times* 11 December 2005.

4. David R. Valee and Michael R. Dion, "Southern N.E. Tropical Storms and Hurricanes, A Ninety-Eight Year Survey 1909–1997." National Weather Service, Taunton, Mass.

5. Valerie Bauerlein, "Hurricane Debate Shatters Civility of Weather Service." *Wall Street Journal* 10 December 2005.

Chapter 2 (pages 21–56)

1. "The Storm That Drowned a City." NOVA transcript, PBS, 2005.
2. Ibid.
3. Ken Wells, "A Retired Oysterman Returns Home." *Wall Street Journal* 29 November 2005.
4. Brian Duffy, "Anatomy of a Disaster." USNews.com, 26 September 2005. www.usnews.com/news/articles/050926/26saturday.htm.
5. Mayor Nagin, quoted in "Hurricane Katrina Timeline." *Wikipedia.* en.wikipedia.org/wiki/Timeline_of_hurricane_katrina.
6. Ibid.
7. Lee Cowan, "New Way of Life in French Quarter." CBSNews.com, 6 September 2005. www.cbsnews.com/stories/2005/09/06/48hours/main820513.shtml?source=RSSattr=48Hours_820513.
8. Ibid.
9. Karen Brooks and Pete Slover, "New Orleans 911 Tapes Reveal Range of Emotions." *Dallas Morning News* 15 September 2005.
10. John McPhee, "The Control of Nature," 23 February 1987. *New Yorker* web site, posted 12 September 2005. www.newyorker.com/archive/2005/09/12/050912ta_talk_McPhee.
11. "Early History." Gateway!NewOrleans.com. www.gatewayno.com/history/LaPurchase.html.
12. Mark Twain, *Life on the Mississippi.* New York: Harper, 1903.
13. John Barry, *Rising Tide: The Great Mississippi Flood of '27 and How It Changed America.* New York: Simon and Schuster, 1997. p. 22.
14. Ibid., p. 127.
15. Ibid., p. 86.
16. Ibid., p. 87.
17. Ibid., p. 221.
18. Ibid., p. 410.
19. Ibid., p. 410.

20. John McPhee, "The Control of Nature."
21. Ibid.
22. John Schwartz, "Levees Are Piece of a $32 Billion Pie." *New York Times* 22 November 2005.
23. Ibid.
24. Charles Mann, "The Long Strange Resurrection of New Orleans." *Fortune Magazine* 29 August 2006. http://money.cnn.com/magazines/fortune_archive/2006/08/21/8383661/index.htm.
25. Ibid.
26. Ibid.
27. Marie Arana, "World Enough and Time." *Washington Post* 30 November 2003, BW10.
28. Mann.
29. Ibid.
30. Shaila Dewan, "Behind Louisiana Aid Package, a Change of Heart by One Man." *New York Times*, 20 March 2006, A1.
31. Mann.
32. Patrick Johnson, "Plan Allows Entire Big Easy to be Rebuilt." *New York Times* 14 April 2006.
33. John Schwartz and Adam Nossiter, "Complex Equation Determined Rules of Rebuilding in New Orleans." *New York Times* 14 April 2006.
34. Mann.
35. Gordon Russell, "Advisories to Raise Houses Also Raising Blood Pressure." *Times Picayune* 14 April 2006; "Citizens Speak Out." *Times Picayune* web site, NOLA.com.
36. Mann.
37. Mike Tidwell, *Bayou Farewell*. New York: Vintage Books, Random House, 2003.

Chapter 3 (pages 57–111)

1. Michael Grunwald. *The Swamp; The Everglades, Florida and the Politics of Paradise*. New York: Simon and Schuster, 2005. p. 187.
2. "Hurricane History." National Hurricane Center NOAA web site. www.nhc.noaa.gov/HAW2/english/history.shtml.
3. Grunwald, p. 105.

4. Florence Fritz, *Unknown Florida*. Coral Gables, Fla.: University of Miami Press, 1964. p. 124.
5. Ibid., p. 120.
6. Grunwald, p. 188.
7. Grunwald, p. 191.
8. "Hurricane History."
9. Michael Grunwald, p. 352.
10. David G. Campbell, *The Ephemeral Islands: A Natural History of the Bahamas*. London: Macmillan, 1978.
11. Marjory Stoneman Douglas, *The Everglades: River of Grass*. Marietta, Ga.: Mockingbird Books, 1974. p. 19.
12. Jonathan Shaw, "Fueling Our Future." *Harvard Magazine* May–June 2006, p. 40.
13. Dauphin Island web site. www.dauphinisland.org.
14. Eduardo Porter, "Damage to the Economy Is Deep and Wide." *New York Times* 31 August 2005, C1.
15. Dauphin Island web site.
16. Fritz, p. 12.
17. "The Real Florida Is a Unique Island." *St. Petersburg Times* 30 October 2005.
18. Necee Regis, "Captiva Comes Back." *Boston Globe* 12 February 2006, M1.
19. Town of Ipswich News Blog. 17 May 2006. www.town.ipswich.ma.us.
20. Abby Goodnough, "Forecasters Predict Active Hurricane Season." *New York Times* 22 May 2006.
21. Valerie Bauerlain, "On Topsail Island: Storms Fuel Battle Over Right to Build." *Wall Street Journal* 7 December 2005.
22. Ibid.
23. "A History of Cape May." Cape May Court Department of Tourism web site. www.thejerseycape.org/historic/county_history.htm.
24. Gilbert Gaul, "A Perpetual Battle with Erosion." *Philadelphia Enquirer* 11 March 2000.
25. "Ash Wednesday Storm 1962." *Wikipedia*. en.wikipedia.org/wiki/Ash_wednesday_storm_of_1962.
26. Al Gore, *An Inconvenient Truth*. Emmaus, Penn.: Rodale, 2006. p. 29.
27. Orrin Pilkey, Interview. Skywatcher Center web site. 21 September 2005. www.earthsky.org/article/48984orrin_pilkey_interview.

28. William Sargent, *Storm Surge*. Lebanon, N.H.: University Press of New England, 2005. p. 106.
29. *Second Skidaway Conference on America's Eroding Shoreline*. Skidaway Institute, Savannah, Georgia. June 1985.
30. Patrick Barry, "Sooner or Later, the Water Will Arrive." *New Scientist* 3 June 2006.
31. *Hurricane of '38*. American Experience PBS web site. www.pbs.org/wgbh/amex/hurricane38.
32. Ibid.
33. Ibid.
34. William Broad, "High Tech Flood Control, With Nature's Help." *New York Times* 6 September 2005.
35. Ibid.
36. Ibid.
37. Molly Moore, "Netherlands Flood Specialists Ponder New Orleans' Plight." *Washington Post* 10 September 2005.
38. Ibid.
39. Patricia Bellis Bixel and Elizabeth Hayes Turner, *Galveston and the 1900 Storm*. Austin: University of Texas Press, 2000. p. 12.
40. Indianola, Texas. *Wikipedia*. http://en.wikipedia.org/wiki/Indianola%2c_texas.
41. Ibid.
42. Bixel and Turner, p. 12.
43. Bixel and Turner, p. 24.
44. Bixel and Turner, p. 33.
45. William Sargent, unpublished manuscript.

Chapter 4 (pages 112–128)

1. Peter Gelling, "As Tsunami Death Toll Nears 400, Indonesians Flee to Hills." *New York Times* 19 July 2006.
2. "Heat Wave in Europe." *Agence France Presse* 18 July 2006.
3. Jennifer Steinhauer, "The Heat Is On." *New York Times* 18 July 2006, A14.
4. Susan Saulny, "Investors Lead Home Sales Boom." *New York Times* 9 July 2006.
5. Susan Saulny, "A Legacy of the Storm: Depression and Suicide." *New York Times* 21 June 2006.

6. Leslie Eaton, "Hurricane Aid Flowing to Homeowners." *New York Times* 17 July 2006.
7. "Monday Night Football." CBS, 24 September 2006.
8. Mark Schleifstein, "Hurricane Katrina and Rita Turned 217 Square Miles of Coastal Land into Water." *Times Picayune* 11 October 2006.
9. Ron Scherer, "New Combatant Against Global Warming: Insurance Industry." *Christian Science Monitor* 14 October 2006.
10. Wendi Goldsmith, personal communication. 25 September 2006.
11. "Dutch Frigate in Biloxi Assists in Hurricane Katrina Efforts." The Royal Netherlands Embassy, Washington, D.C., web site. 7 September 2005. www.netherlands-embassy.org/article.asp?articleref =AR00001756EN.
12. Ibid.
13. Ibid.
14. Len Bahr, personal communication. 26 September 2006.
15. Nancy Anne Newhouse, *Voices of the Village: An Oral History of Sconset.* Sconset, Mass.: Sconset Trust, Inc., 2004.
16. Peter B. Brace, "Sankaty Light to Be Moved in 2007." *Nantucket Independent* 26 July 2006.
17. Adam Nossiter, "Blakeley's Style Confounds and Comforts." *New York Times* 16 April 2007.

Bibliography

Katrina was probably the most thoroughly covered natural disaster of all time. This book draws on many sources, including personal interviews, telephone calls, newspaper articles, scientific papers, legal documents, and television broadcasts. In the Notes section I have tried to include the many sources that helped tell the story and shaped my thinking. Here I would like to highlight some of the books that were particularly helpful. I hope they will be of use to others who may want to explore these topics further.

Certainly the two most colorful books that cover the history of the Mississippi are Mark Twain's classic, *Life on the Mississippi* (James R. Osgood and Co., 1883), and James Barry's *Rising Tide: The Great Mississippi Flood of '27 and How It Changed America* (Simon and Schuster, 1997). It is encouraging that Mr. Barry now heads the newly reorganized Levee Board in New Orleans.

Orrin Pilkey wrote *The Beaches Are Moving* (Doubleday, 1979), the seminal book that espouses coastal retreat. Its arguments have been updated in *Against the Tide*, by Cornelia Dean (Columbia University Press, 1999). Both books evolved naturally from Arthur Strahler's classic, *A Geologist's View of Cape Cod* (Natural History Press, 1996).

Robert Penn Warren was awarded a Pulitzer prize for *All the*

King's Men (Modern Library, 1946), which tells the lightly fiction-alized story of Louisiana politics under Governor Huey Long. Douglas Brinkley has written the most critical and detailed history of Katrina in *The Great Deluge* (William Morrow, 2006). Chris Rose, former entertainment critic for the *Picayune Times*, found his voice in the aftermath of Katrina and wrote the poignant and amusing *1 Dead in Attic* (Chris Rose Books, 2005). Mike Tidwell wrote the evocative *Bayou Farewell* (Vintage, 2003) about the destruction of Louisiana's coastlines and Cajun way of life. John McQuaid and Mark Schleifstein, also of the *Picayune Times*, wrote *Path of Destruction* (Little, Brown, 2006), which describes the devastation of New Orleans and our present age of superstorms. John Miller edited *New Orleans Stories* (Chronicle Books, 1992), a fascinating collection of writings by some of the Crescent City's favorite sons and daughters, including Louis Armstrong, James Audubon, Truman Capote, and Anne Rice.

Marjory Stoneman Douglas wrote *The Everglades: River of Grass* (Mockingbird Books, 1947), a classic natural history book that ignited the movement to save this national treasure. Michael Grunwald updated that movement in *The Swamp: The Everglades, Florida and the Politics of Paradise* (Simon and Schuster, 2006). Florence Fritz unearthed fascinating stories about Florida's past, particularly along the colorful Gulf Coast, in *Unknown Florida* (University of Miami Press, 1963).

Of course, Al Gore has written the highly influential book *An Inconvenient Truth* (Rodale, 2006), which lays out the details of global warming in compelling, accessible language. Its conclusions have been backed up by the recent reports by the UN panel on climate change. It is unfortunate that *An Inconvenient Truth* did not come out thirty years earlier, when we still had a fighting chance to do something substantial to reduce global warming. Now we are in the unenviable position of having to simply adapt to the many changes caused by more intense storms and sea-level rise.

Patricia Bixel and Elizabeth Turner wrote *Galveston and the 1900 Storm* (University of Texas Press, 2000), which chronicles the destruction caused by the nation's deadliest hurricane and the valiant recovery of this southern city. Erik Larson chronicled the fledgling days of hurricane forecasting in his well-researched book *Isaac's Storm* (Random House, 2000).

Index